Disclaimer

Book Title: Experimental Characterization of Helium Dispersion in a -Scale Two-Car Residential Garage (NIST TN 1694)

Book Author: William M. Pitts; Jiann C. Yang; Marco G. Fernandez;

Book Abstract: A series of experiments are described in which helium was released at constant rates into a 1.5 m 1.5 m 0.75 m enclosure designed as a -scale model of a two car garage. The purpose was to provide reference data sets for testing and validating computational fluid dynamics (CFD) models and to experimentally characterize the effects of a number of variables on the mixing behavior within an enclosure and the exchange of helium with the surroundings. Helium was used as a surrogate for hydrogen, and the total volume released was scaled as the amount that could be released by a typical hydrogen-fueled automobile with a full fuel tank. Temporal profiles of helium were measured at seven vertical locations within the enclosure during and following one hour and four hour releases. Idealized vents in one wall sized to provide air exchange rates typical of actual garages were used. The effects of vent size, number, and location were investigated using three different vent combinations. The dependence on leak location was considered by releasing helium from three different points within the enclosure. A number of tabulated quantitative measures are used to characterize the experiments. The complete experimental measurement results for each condition are available on the internet as described in Appendix A (http://www.nist.gov/el/fire_protection/buildings/upload/HeliumDispersionData Sets.zip.)

Citation: NIST TN - 1694

Keywords: concentration measurements, differential pressure, doorway fan tests, garages, helium mixing, hydrogen, hydrogen-fueled automobiles, reference data sets

NIST Technical Note 1694

Experimental Characterization of Helium Dispersion in a ¼-Scale Two-Car Residential Garage

William M. Pitts
Jiann C. Yang
Marco G. Fernandez

NIST

National Institute of Standards and Technology • U.S. Department of Commerce

NIST Technical Note 1694

Experimental Characterization of Helium Dispersion in a ¼-Scale Two-Car Residential Garage

William M. Pitts
Jiann C. Yang
Marco G. Fernandez
Engineering Laboratory
Fire Research Division

March 2011

U.S. Department of Commerce
Gary Locke, Secretary

National Institute of Standards and Technology
Patrick D. Gallagher, Director

National Institute of Standards and Technology Technical Note 1694
Natl. Inst. Stand. Technol. Tech. Note 1694, 85 pages (March 2011)
CODEN: NSPUE2

Abstract

A series of experiments are described in which helium was released at constant rates into a 1.5 m × 1.5 m × 0.75 m enclosure designed as a ¼-scale model of a two car garage. The purpose was to provide reference data sets for testing and validating computational fluid dynamics (CFD) models and to experimentally characterize the effects of a number of variables on the mixing behavior within an enclosure and the exchange of helium with the surroundings. Helium was used as a surrogate for hydrogen, and the total volume released was scaled as the amount that could be released by a typical hydrogen-fueled automobile with a full fuel tank. Temporal profiles of helium were measured at seven vertical locations within the enclosure during and following 1 h and 4 h releases. Idealized vents in one wall sized to provide air exchange rates typical of actual garages were used. The effects of vent size, number, and location were investigated using three different vent combinations. The dependence on leak location was considered by releasing helium from three different points within the enclosure. A number of tabulated quantitative measures are used to characterize the experiments. The complete experimental measurement results for each condition are available on the internet as described in Appendix A.

Keywords: concentration measurements, differential pressure, doorway fan tests, garages, helium mixing, hydrogen, hydrogen-fueled automobiles, reference data sets

Table of Contents

Abstract i
Table of Contents...ii
List of Tables iii
List of Figures iv
1.   Introduction 1
     1.1.   Background..1
     1.2.   Previous Work 1
     1.3.   Problem Description 4
2.   Experimental..5
     2.1.   Reduced-Scale Garage 5
     2.2.   Helium Sensor Calibration and Concentration Measurement......................................7
     2.3.   Pressure Measurement 17
     2.4.   "Doorway Fan Test" 19
     2.5.   Time-Resolved Helium Volume Fraction Measurements in the Reduced-Scale Garage 19
3.   Experimental Results 20
     3.1.   Fan Test Characterization of Vent Configurations 20
     3.2.   Time-Resolved Helium Concentration and Pressure Measurements in the Reduced-Scale
            Garage 22
          3.2.1. One Hour Helium Releases near the Floor at the Center of the Garage 22
          3.2.2. One Hour Helium Releases near the Floor at the Rear of the Garage 38
          3.2.3. One Hour Helium Releases near the Ceiling at the Center of the Garage.............42
          3.2.4. Four Hour Helium Releases near the Floor at the Center of the Garage 48
          3.2.5. Four Hour Helium Releases near the Floor at the Rear of the Garage 54
          3.2.6. Four Hour Helium Releases near the Ceiling at the Center of the Garage 58
4.   Discussion...63
5.   Final Remarks 69
6.   References 69
Appendix A   Downloadable Data Files of Experimental Results for the Eighteen Tests......................75
http://www.nist.gov/el/fire_protection/buildings/upload/HeliumDispersionDataSets.zip

List of Tables

Table 1.   Heights for Vertical Array of Seven Helium Sensors Located 37.5 cm from the Front and Side Walls of the Reduced-Scale Garage 19

Table 2.   Front Wall Vent Characteristics 22

Table 3.   Summary of Quantitative Measures for Eighteen Test Conditions 26

Table 4.   Variations in Differential Pressure Associated with Changes in Flow Conditions (ΔP)........32

List of Figures

Figure 1.   A schematic and photograph of the ¼-scale two car garage are shown. 5

Figure 2.   The measured output voltage for a Neodym Technologies thermal conductivity sensor is shown as a function of time as the helium volume fraction is stepped over a 0 to 1 range. 8

Figure 3.   An example of the fourth order polynomial fit to experimental data for the response of a Neodym Technologies sensor to varying helium volume fractions. 9

Figure 4.   Photograph of a TCG-3880 showing the associated mount, external heater resistor, and wiring. 10

Figure 5.   The measured output voltage for one of the Xensor Integration thermal conductivity sensors is shown as a function of time as the helium volume fraction is stepped over a 0 to 0.2 range. 11

Figure 6.   An example of a fourth order polynomial fit to experimental data for the response of a Xensor Integration sensor to varying helium volume fractions. 11

Figure 7.   Four repeated calibrations over a five-month period of the response of a Xensor Integration probe to helium volume fractions ranging from 0 to 1 are shown. 12

Figure 8.   A portion of the data shown in Figure 5 is blown up in order to emphasize the sensor time response. 13

Figure 9.   The percentage of helium recorded by Xensor Integration sensors over a roughly two and a half day period at eight locations within a sealed enclosure are shown. 14

Figure 10. Helium volume fraction measured with Xensor Integration sensors is plotted as a function of time for eight locations within the enclosure which was sealed and unsealed multiple times. A flow of air was introduced into the enclosure at around 53 hours. 14

Figure 11. The data shown in Figure 10 is replotted on expanded percentage helium and time scales ...15

Figure 12   Average helium volume percents measured by eight Xensor Integration sensors are plotted against the corresponding average for all of the sensors for a range of helium concentration in the reduced-scale garage. 16

Figure 13. Experimental and predicted voltages for the response of a Xensor Integration sensor to a range of helium concentrations is shown. 17

Figure 14   The difference between experimental values of helium volume fraction and values calculated using the approach described in the text is plotted as a function of the nominal helium volume percent. 18

Figure 15. The results for three calibrations of the Baratron electronic manometer are plotted as differential pressure versus voltage. The straight line is a linear least squares curve fit to the data..18

Figure 16. Values of air flow rate are plotted against the measured differential pressure for three repeated fan tests of the reduced-scale garage with a single 2.4 cm × 2.4 cm opening in the front wall. The solid line is the result of a non-linear least squares curve fit to Eq. (2). 21

Figure 17. Values of air flow rate are plotted against the measured differential pressure for three repeated fan tests for the reduced-scale garage with a single 3.05 cm × 3.05 cm opening in the front wall. The solid line is the result of a non-linear least squares curve fit to Eq. (2). 21

Figure 18. Values of air flow rate are plotted against the measured differential pressure for three repeated fan tests for the reduced-scale garage with two 2.15 cm × 2.15 cm opening in the front wall. The solid line is the result of a non-linear least squares curve fit to Eq. (2). 22

Figure 19. The upper plot shows helium volume percent as a function of time at eight locations for a one hour release of helium from the lower center position into the ¼-scale garage with a single centered 2.4 cm × 2.4 cm opening in the front face. The lower plot shows the differential pressure across the face..23

Figure 20. Helium volume percent measurements recorded at eight locations within the ¼-scale garage equipped with a single 2.4 cm × 2.4 cm vent are shown for the initial period of a one hour helium release near the floor at the center of the garage. 24

Figure 21. Helium volume percent measurements recorded at eight locations within the ¼-scale garage equipped with a single 2.4 cm × 2.4 cm vent are shown for a period ranging from 200 s before to 2800 s after the flow was halted for a one hour helium release near the floor at the center of the garage. 25

Figure 22. The differential pressure is shown for 3600-LC-SSV over the time from just before to just after a 29.6 L/min air flow was used to sweep remaining helium from the garage. 32

Figure 23. Helium volume percent measurements recorded at eight locations within the ¼-scale garage equipped with a single 3.05 cm × 3.05 cm vent are shown for the initial period of a one hour helium release near the floor at the center of the garage. 33

Figure 24. The upper plot shows helium volume percent as a function of time at eight locations for a one hour release of helium from the lower center position into the ¼-scale garage with 2.15 cm × 2.15 cm upper and lower openings in the front face. The lower plot shows the differential pressure across the face..34

Figure 25. Helium volume percent measurements recorded at eight locations within the ¼-scale garage equipped with upper and lower 2.15 cm × 2.15 cm vents are shown for the initial period of a one hour helium release near the floor at the center of the garage. 35

Figure 26. The four plots show height above the floor as a function of helium volume percent immediately following the end of helium release for 3600-LC-ULV (upper left), the corresponding density plot (upper right), relative vertical hydrostatic pressure difference distributions inside the garage and in the ambient (lower left), and the calculated vertical differential pressure distribution with the neutral plane located to provide the required differential pressure ratio at the vents (lower right). 37

Figure 27. The upper plot shows helium volume percent as a function of time at eight locations for a one hour release of helium from the lower rear position into the ¼-scale garage with a single centered 2.4 cm × 2.4 cm opening in the front face. The lower plot shows the differential pressure across the face. 39

Figure 28. The upper plot shows helium volume percent as a function of time at eight locations for a one hour release of helium from the lower rear position into the ¼-scale garage with a single centered 3.05 cm × 3.05 cm opening in the front face. The lower plot shows the differential pressure across the face. Data acquisition stopped unexpectedly at 5808 s........40

Figure 29. The upper plot shows helium volume percent as a function of time at eight locations for a one hour release of helium from the lower rear position into the ¼-scale garage with 2.15 cm × 2.15 cm upper and lower openings in the front face. The lower plot shows the differential pressure across the face. 41

Figure 30. The upper plot shows helium volume percent as a function of time at eight locations for a one hour release of helium from the upper center position into the ¼-scale garage with a single centered 2.4 cm × 2.4 cm opening in the front face. The lower plot shows the differential pressure across the face. 43

Figure 31. The upper plot shows helium volume percent as a function of time at eight locations for a one hour release of helium from the upper center position into the ¼-scale garage with a single centered 3.05 cm × 3.05 cm opening in the front face. The lower plot shows the differential pressure across the face. 45

Figure 32. The upper plot shows helium volume percent as a function of time at eight locations for a one hour release of helium from the upper center position into the ¼-scale garage with 2.15 cm × 2.15 cm upper and lower openings in the front face. The lower plot shows the differential pressure across the face. 46

Figure 33. Values of experimental helium volume percents (solid symbols) are plotted as function of distance above the floor for 3600-LC-ULV (circle) and 3600-UC-ULV (square). Estimated values at the floor and ceiling are indicated by open symbols..............................47

Figure 34. The upper plot shows helium volume percent as a function of time at eight locations for a four hour release of helium from the lower center position into the ¼-scale garage with a single centered 2.4 cm × 2.4 cm opening in the front face. The lower plot shows the differential pressure across the face. 48

Figure 35. Values of experimental helium volume percents (solid symbols) are plotted as a function of distance above the floor for 3600-LC-SSV (circle) and 14400-LC-SSV (square). Estimated values at the floor and ceiling are indicated by open symbols..............................49

Figure 36. The upper plot shows helium volume percent as a function of time at eight locations for a four hour release of helium from the lower center position into the ¼-scale garage with a single centered 3.05 cm × 3.05 cm opening in the front face. The lower plot shows the differential pressure across the face...51

Figure 37. Values of experimental helium volume percents (solid symbols) are plotted as a function of distance above the floor for 3600-LC-SLV (circle) and 14400-LC-SLV (square). Estimated values at the floor and ceiling are indicated by open symbols..............................52

Figure 38. The upper plot shows helium volume percent as a function of time at eight locations for a four hour release of helium from the lower center position into the ¼-scale garage with 2.15 cm × 2.15 cm upper and lower openings in the front face. The lower plot shows the differential pressure across the face...53

Figure 39. The upper plot shows helium volume percent as a function of time at eight locations for a four hour release of helium from the lower rear position into the ¼-scale garage with a single centered 2.4 cm × 2.4 cm opening in the front face. The lower plot shows the differential pressure across the face...54

Figure 40. The upper plot shows helium volume percent as a function of time at eight locations for a four hour release of helium from the lower rear position into the ¼-scale garage with a single centered 3.05 cm × 3.05 cm opening in the front face. The lower plot shows the differential pressure across the face...56

Figure 41. The upper plot shows helium volume percent as a function of time at eight locations for a four hour release of helium from the lower rear position into the ¼-scale garage with 2.15 cm × 2.15 cm upper and lower openings in the front face. The lower plot shows the differential pressure across the face...57

Figure 42. The upper plot shows helium volume percent as a function of time at eight locations for a four hour release of helium from the upper center position into the ¼-scale garage with a single centered 2.4 cm × 2.4 cm opening in the front face. The lower plot shows the differential pressure across the face...58

Figure 43. Values of experimental helium volume percents (solid symbols) are plotted as a function of distance above the floor for 14400-LC-SSV (circle) and 14400-UC-SSV (square). Estimated values at the floor and ceiling are indicated by open symbols..............................59

Figure 44. The upper plot shows helium volume percent as a function of time at eight locations for a four hour release of helium from the upper center position into the ¼-scale garage with a single centered 3.05 cm × 3.05 cm opening in the front face. The lower plot shows the differential pressure across the face...60

Figure 45. The upper plot shows helium volume percent as a function of time at eight locations for a four hour release of helium from the upper center position into the ¼-scale garage with 2.15 cm × 2.15 cm upper and lower openings in the front face. The lower plot shows the differential pressure across the face...62

Figure 46. Values of experimental helium volume percents (solid symbols) are plotted as a function of distance above the floor for 14400-LC-ULV (circle) and 14400-UC-ULV (square). Estimated values at the floor and ceiling are indicated by open symbols..............................63

1. Introduction

1.1. Background

Concerns about climate change are driving efforts to develop hydrogen-powered systems as replacements for many current applications utilizing hydrocarbon fuels. A number of demonstrations are underway that are designed to show that hydrogen can be used for mobile and stationary applications. The ultimate goal is to develop a hydrogen-based economy.

The physical properties of hydrogen differ from those for hydrocarbon fuels. As a result, the mixing and combustion behaviors of hydrogen differ in significant ways from hydrocarbons and must be taken into account when engineering systems for safe operation and fire prevention. Efforts are underway to develop standards and codes appropriate for hydrogen-fueled systems. The differences between hydrogen and typical hydrocarbon fuels are particularly important when hydrogen is released into enclosed spaces such as building, garages, and tunnels. Example applications include hydrogen-fueled automobiles parked in residential garages and stationary fuel cells located within a building.

1.2. Previous Work

The unique fire safety problems associated with hydrogen have led to a number of studies aimed at experimentally characterizing the temporal mixing behaviors of hydrogen releases within enclosures as well as the application of flow models for predicting these behaviors. One of the earliest studies was performed by Koontz et al. who measured the temporal behavior of hydrogen concentrations at six locations within a two-car garage following hydrogen release near the floor or generation during the charging of a battery. [1] Swain et al. used a computational fluid dynamics (CFD) code to simulate these experiments and validate the modeling. [2] This group had applied CFD modeling to the problem of hydrogen release in enclosures previous to this time. [3]

Many researchers have been reluctant to use hydrogen for experimental testing due to safety concerns. It has been common to use helium instead as a surrogate. The density of helium is twice as large, and its molecular diffusion coefficient is roughly 90 % of hydrogen. Swain et al. investigated the differences between helium and hydrogen releases using a CFD approach validated by comparison with helium measurements. [4] Time-resolved helium concentration measurements in a ½-scale corridor equipped with various types of vents in which helium was released were shown to agree well with CFD predictions. A formalized approach based on validating a CFD model using helium volume fraction measurements within a partially sealed volume and then using the model to predict hydrogen concentrations and the related risk was described by Swain et al. in a separate publication [5] Agranat et al. also used a portion of this data to validate a CFD approach to modeling this type of flow. [6] They then used the code to study hydrogen mixing and transport in enclosed refueling stations. Swain et al. later provided more detailed comparisons of predicted time-resolved helium and hydrogen distributions during releases into enclosures. [7]

In an unpublished contractor report Swain described time-resolved helium measurements at four locations within a full-scale single car garage. [8] The helium was released from underneath a wooden mock-up of an automobile. Limited CFD modeling of the helium releases agreed well with the experimental measurements. Papanikolaou and Venetsanos used a CFD model to simulate the Swain results in more detail. [9] In 2001 Breitung et al. reported a CFD analysis of short period releases of hydrogen from an automobile parked inside a single-car garage with two vents near the ceiling. [10]

A variety of CFD codes for modeling hydrogen dispersion were tested in a round robin study involving twelve laboratories [11] that simulated hydrogen concentrations measured by Shebeko et al. at six locations along the vertical direction of a large sealed cylindrical vessel [12]. There were substantial variations in calculated hydrogen concentrations at various times for the different models. The calculated values generally clustered about the experimental values within a range of values from 0.5 to 2 times the experimental values. In a few cases, the calculated values were more than factors of 10 times lower. An earlier CFD study by Puzack used the same experimental data to validate the modeling. [13]

During the Second International Conference on Hydrogen Safety, a number of papers reported experimental findings on the release and dispersion of helium [14] or hydrogen [15,16,17] in real-scale enclosures representative of garages. Note that some of these papers have subsequently been published, and, when available, the published citations have been provided. Gupta et al. [14] and Lacome et al. [15] utilized enclosures with small openings near the base to minimize pressure differences between the inside and outside of the enclosure, while limiting the loss of injected gas. Both groups reported real-time concentration measurements at multiple locations. Injected gas volume flow rates and durations were varied. Tchouvelev et al. studied hydrogen released into an enclosure containing a fan for mixing. [16] Ishimoto et al. considered releases of hydrogen into a ventilated enclosure. [17]

One of the experiments described by Lacome et al. [15] served as the basis for a second round robin study of CFD capabilities. [18] The calculations were run prior to the availability of the experimental results, i.e., a blind test, and then repeated afterwards. As found earlier for a different experimental configuration [11], there were significant differences between the experimental hydrogen concentration measurements and the CFD predicted values at various times. The magnitudes of the differences were similar, with calculated values generally covering a range of 0.5 to 2 of the experimental values. For the worst case in the blind test, outliers spanned a range of 0 to 2.7. The largest differences were for sensors located in the lower part of the enclosure and along the centerline of the buoyant flow. The gaps between the experimental and calculated concentrations were reduced somewhat for calculations made after the experimental data were available, but significant differences remained.

Zhang et al. have described the application of one of the CFD models included in the round robin [18] to the Lacome et al. data [15] in separate manuscripts. [19,20] The agreement between the CFD and experimental results appears to be somewhat better than indicated in the round robin paper. Tchouvelev et al. applied CFD calculations to their measurements of hydrogen in a chamber with fan-assisted mixing. [16] The temporal variations in hydrogen concentrations at the various measurement locations were captured well.

Barley et al. described CFD modeling of hydrogen dispersion in real-scale garages with a goal of understanding and assessing the effectiveness of passive ventilation openings for removing released hydrogen. [21] They found that two openings near the top and bottom of the enclosure were most effective due to the resulting buoyancy-induced flow. Barley et al. also described a simple analytical model for predicting flow in and out of the enclosure based on the hydrostatic pressure differences across the enclosure boundary due to the low-density hydrogen/air mixture and included a factor to capture the effects of stratification. The analytical model agreed well with the CFD predictions for reasonable values of the stratification factor. A later paper from the same group compared experimental measurements with both CFD modeling and the analytical model. [22] Zhang et al. also discussed an analytical approach for predicting hydrogen distributions in an enclosure based on induced hydrostatic pressure differences. [19] Their model was modified from a two-zone model commonly used for enclosure fires that assumes an upper layer of uniformly mixed gases and a lower air layer. Significant differences between predictions and experimental results were attributed to concentration gradients present in the upper layer of the experiment.

Related work dealing with low-speed releases of hydrogen/natural gas mixtures in an enclosure was described by Lowesmith et al. [23]. These authors made concentration measurements at various locations within a real-scale enclosure and developed a simple model for the mixing that yielded predictions that agreed well with the experimental measurements.

Numerous works dealing with the topic of hydrogen (or helium release) in enclosures were also presented during the Third International Conference on Hydrogen Safety. Three of these papers considered the hydrogen distributions expected when hydrogen permeates very slowly through fuel tank walls into an enclosure having very small openings. [24,25,26] Very recently, revised publications based on the same work have appeared. [27,28,29,30] A manuscript by Cariteau et al. described measurements in which helium was released into a full-scale garage which was either empty or contained a vehicle. [31] These authors focused on the interior helium distributions resulting from moderately sized helium volume flow rates. Work describing the distribution of hydrogen following releases inside a full-scale garage

2

with and without a vehicle was reported by Merilo et al. [32,33] These authors also studied the effects of igniting the hydrogen-air mixtures. Benteboula et al. considered the effectiveness of two simple models for predicting helium mixing behavior for a series of experiments carried out in the same facility described by Gupta et al. [14]. [34] Denisenko et al. investigated the effects of release conditions on the mixing of helium inside a real-scale enclosure. [35] Their findings demonstrated that low-speed releases which were buoyancy-dominated resulted in the formation of a stratified upper layer, while high-speed, momentum-dominated releases resulted in mixing over the entire enclosure and created homogeneous mixtures. A model for the deflagration of hydrogen-air mixtures within enclosures was described by Skob et al. [36]

The third in a series of round-robin studies in which a number of CFD codes were applied to predict the older experimental findings of Swain [8] for mixing in a garage for a helium release in the presence of a vehicle was described at the Symposium and has appeared in print. [37,38] This round robin involved four organizations. Similar to the two earlier round robins [11,18], there were substantial variations between the model predictions and experimental results. This paper is noteworthy because it utilized numerical measures for assessing the degree of agreement between models and experiment and emphasized the potential role of experimental uncertainty when comparing modeling and experimental findings.

Three manuscripts, subsequently published, dealt with unintended releases of hydrogen into fuel cell enclosures. The first by Friedrich et al. describes experiments characterizing hydrogen distributions and combustion behavior following the release of hydrogen into the interior of a simulated fuel-cell module. [39] The second experiment was similar but incorporated a second enclosure, representing a room, around the simulated fuel-cell cabinet. [40] The third report described an experimental and modeling study designed to determine minimum venting required for an enclosure housing a fuel cell system in order to avoid the build-up of a flammable hydrogen mixture. [41]

A short review of many of the experiments described above was presented during the Third Symposium and subsequently published. [42]

A preliminary version of the work which is the subject of the current manuscript was also presented during the Third Symposium. [43]

A number of other modeling studies dealing with hydrogen release inside enclosures have been published. Middha et al. performed a number of CFD calculations for a variety of configurations and validated the results by comparison with experiments. [44] In a series of papers Matsuura et al. have applied CFD to computationally investigate the role of passive and active venting in removing hydrogen from various configurations of enclosed spaces. [45,46,47,48] Reference [45] is especially relevant because it models unpublished data reported by Swain et al. for a corridor configuration with a vent at the end and in the ceiling. [49] The paper also includes measurements that were apparently made for the same experimental configuration in which real-time concentration probes were utilized [50], in contrast to the measurements of Swain et al., which were widely spaced in time. The remaining papers deal primarily with using CFD for sensing-based risk mitigation. A similar CFD study has been reported by Liu and Schreiber, who computationally investigated the distribution of hydrogen inside an automobile due to an unintended release. [51] Kim et al. have reported CFD modeling of hydrogen distributions in fuel cell enclosures [52] as well as a Lagrangian approach for modeling hydrogen releases in enclosures [53]. Vudumu and Koylu developed another model for buoyancy-dominated flows and applied it to hydrogen releases into a simple cylindrical cylinder. [54] Kanayama and coworkers have published a series of studies dealing with the development of the upper layer when a buoyant flow strikes the ceiling within an enclosed space. [55,56,57] Prasad et al. used preliminary results of the work described here to validate the use of a large-eddy CFD code that was developed to predict fire behavior, FDS [58], for predicting helium and hydrogen dispersion within partially enclosed spaces. [59,60] An analytical model that assumes rapid mixing for predicting helium or hydrogen concentration in one or more layers within an enclosure with multiple vents has been described by Prasad et al. [61]

1.3.   Problem Description

As hydrogen-powered automobiles come on the market, it is likely that they will be parked in the existing stock of residential garages. For this reason, it is important to understand the implications of potential hydrogen leaks in typical residential garages. Even though the literature described above is focused on garages, many of the studies have not fully investigated parameters likely to be important in actual garages. Two such parameters are hydrogen leak location and the size and spatial distribution of leaks. Investigations focused on losses of hydrogen from an enclosure during and following a release are limited.

Leak rates for garages are typically described in terms of the number of air changes per hour (ACH) for an enclosure of volume, V_{enc}, which corresponds to a volume flow exchange rate in m^3/s across the enclosure boundary, Q_{enc}, given by $Q_{enc} = V_{enc} \times ACH/3600$. ACH can vary substantially with time and depends not only on the areas of openings connecting across the enclosure boundary, but also on such factors as weather conditions and forced ventilation which control the nominal pressure differences across openings. Values of Q_{enc} can be related to an effective leak area, ELA, by use of the Bernoulli equation,

$$ELA_H = (Q_{enc})_H /(2 \times \Delta P / \rho)^{\frac{1}{2}}, \tag{1}$$

where the subscript H indicates evaluation at a specific pressure difference between the interior and exterior and gas density, ρ. A value of $H = 4$ Pa is often taken to be representative of the pressure difference across a garage boundary. [62] Note that ELA_H does not normally correspond to the actual open area in a garage boundary since experimental flow rates are typically [62] given by

$$Q_{enc} = C \times \Delta P^n, \tag{2}$$

where C is the flow parameter and n is an experimental parameter that varies between 0.5 and 1.

Studies indicate that values of ACH and $ELA_{4\,Pa}$ vary widely for garages in the United States, e.g., see [62,63]. For testing purposes, it is reasonable to consider ACHs on the low side of those observed. For this purpose, a recommended minimum value of $Q_{enc} = 2.73$ m^3/min (100 ft^3/min) per stored automobile included in an early version of an American Society of Heating, Refrigerating and Air-Conditioning Engineers (ASHRAE) standard was used. [64] Swain et al. referred to this value in an early study. [2] Note that this recommendation is no longer included in current versions of the ASHRAE standard, but is incorporated in the 2009 International Mechanical Code published by the International Code Council. [65] Both codes assume natural ventilation processes are sufficient and do not specify means for meeting the standard. A simple calculation reveals that this value corresponds exactly to an $ACH = 3$ h^{-1} for a single car garage sized 3.048 m (w) × 6.096 m (l) × 3.048 m (h) (10 ft × 20 ft × 10 ft). On this basis, an $ACH = 3$ h^{-1} was adopted as a representative value for this study.

A study published by researchers in Europe after the experiments discussed here were completed concluded that ACH values for garages in Europe and the United States are considerably lower than indicated above. [27] After eliminating results from a Canadian study, which they considered to be unreasonably high, they concluded that experimental measurements of ACHs for residential garages were all less than $ACH = 3$ h^{-1}, and roughly 85 % were less than $ACH = 1$ h^{-1}. As a worst-case example, they chose a value of $ACH = 0.03$ h^{-1}, which is 100 times smaller than the value considered here. It is difficult to imagine that 100 % of garages fail to meet the older ASHRAE or current ICC standards. The source of the large difference between the ACH values discussed here and those cited in [27] is currently unclear, but may be related to the assumption of a nominal 4 Pa pressure difference between the garage interior and surroundings. The source of the difference should be explored further since the effective ventilation rate has a dramatic impact on hydrogen build up in a garage. In any case, it would appear that the low value adopted in [27] is exceedingly conservative.

The purpose of the current study is twofold. The first is to provide a set of reference data for testing and validating the capability of Fire Dynamics Simulator (FDS) [58], a code developed at the National Institute of Standards and Technology (NIST) for simulating buoyancy-dominated fire flows,

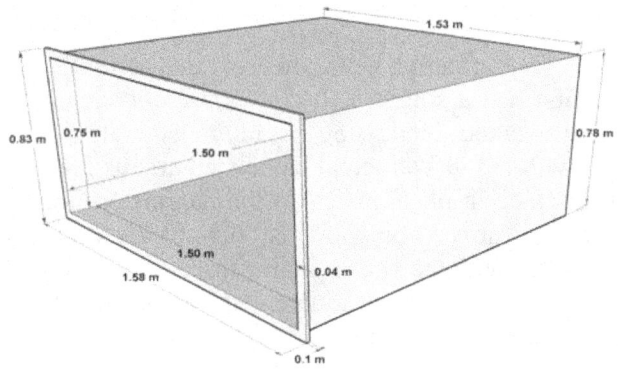 

Figure 1.   A schematic and photograph of the ¼-scale two-car garage are shown.

and other computational fluid dynamic (CFD) codes to predict concentration distributions within an enclosure, including losses to the ambient surroundings, of a buoyant gas during release and post-release periods. The second is to experimentally characterize the effects of the gas release point, the buoyant gas release rate, and the vent size(s) and location(s) on the mixing behavior and interior concentration profiles during the release and post-release periods.

For modeling purposes it is desirable to have accurate, repeatable experiments with well defined boundary conditions and initial conditions. Such control is exceedingly difficult in actual garages, which are subject to outside weather conditions (i.e., changing winds and temperatures) and have leaks that are difficult to characterize. The choice was made to perform measurements in a well-controlled laboratory environment. In order to maintain the test facility at a manageable size and limit the amount of buoyant gas required, a scaled enclosure was used. Simple vent configurations, single and double vents located in a single wall, were used for the initial measurements.

It should be noted that the reduced-scale experiments were not designed to provide a full similitude model of a full-scale garage. Due to the nature of the system, it was not possible to match all of the dimensionless numbers expected to be important. The experiment is designed such that spatial scales, gas volume flow rates, and flow times are scaled to match those corresponding to a full-scale experiment.

2.   Experimental

2.1.   Reduced-Scale Garage

The reduced-scale experiment is based on the following highly idealized scenario. A release of 5 kg of hydrogen (representative of a full tank on current designs of hydrogen-powered automobiles) occurs at room temperature within a two-car garage having interior dimensions of 6.096 m × 6.096 m × 3.048 m. The hydrogen is completely released at a constant rate over one hour or four hours at a fixed location within the garage.

A physical scale model having interior dimensions of 1.5 m × 1.5 m × 0.75 m was constructed from nominally 1.27 cm thick poly(methyl methacrylate) (PMMA) sheets. The corresponding scaling factor is 0.246. Five PMMA sheets were glued together to form a sealed enclosure with an open front end. A 2.54 cm wide flange was attached around the outside edge of the open end, and a 1.576 m × 0.818 m removable front face was placed over the flange and held in place with a series of clamps. A polychloroprene gasket and stopcock grease were used to form a tight seal between the front face and flange. Figure 1 shows a schematic for the enclosure along with a photograph. The integrity of the enclosure was confirmed by introducing helium into the enclosure and testing for leaks by probing

5

outside of the enclosure using a Varian Model 959 helium leak detector.[1] Several small leaks were sealed with silicon sealant.

All vents, a pressure tap, and electrical and gas feedthroughs were incorporated into the removable front faces. Three faces were fabricated. The first had a single 2.40 cm ± 0.02 cm square (5.76 cm^2 area) opening at the center of the face. This size was determined using Eq. (1) to estimate the vent area required for $ACH = 3$ h^{-1} ($Q_{enc} = 1.41 \times 10^{-3}$ m^3/s) with $H = 4$ Pa (actual calculated area is 5.46 cm^2). Subsequent measurements (described below) with this face in place provided the parameters for use in Eq. (2). The calculated value of $(Q_{enc})_{4\ Pa}$ was 9.29×10^{-4} m^3/s. This value is 66 % of the value based on Eq. (1). Based on this result, a second face was prepared with the vent area increased by a factor of 1.60, i.e., a square with 3.05 cm ± 0.02 cm sides (9.30 cm^2 area). Two equal square vents with 2.15 cm ± 0.02 cm sides (total area = 9.25 cm^2) were placed in the third face equal distances from the sidewalls with the bottom edge of the lower 2.54 cm above the floor and the top edge of the upper 2.54 cm below the ceiling. Experimental measurements with the latter two front faces (described below) showed that the predicted values of $(Q_{enc})_{4\ Pa}$ were near the value of 1.41×10^{-3} m^3/s necessary for $ACH = 3$ h^{-1}.

Due to safety concerns, helium was chosen as the buoyant gas for this study. A mass flow controller was used to deliver a constant volume flow rate chosen such that the total volume of helium delivered during the release period (either 3600 s or 14 400 s) corresponded to the dimensionally scaled volume (59.8 m^3) for a release of 5 kg of hydrogen into the full-scale garage. The corresponding volume of 0.890 m^3 for the reduced-scale garage gives flow volume rates of 2.47×10^{-4} m^3/s = 14.8 L/min and 6.18×10^{-5} m^3/s = 3.71 L/min for the one hour and four hour releases, respectively. The actual volume flow rates delivered by the mass flow controller were measured to be 14.92 L/min ± 0.15 L/min and 3.54 L/min ± 0.06 L/min using a Gilabrator-2 electronic bubble flow meter from Gilian. The uncertainties listed are based on standard deviations from repeated measurements spanning the several month period required to complete the testing. In an unpublished report, Mulholland and Fernandez found the expanded uncertainty (2σ) for this instrument to be 1.8 %. [66]

The helium flowed by way of tubing and a feedthrough into the enclosure where it was released from a Fischer burner, chosen as a convenient means to release helium over an area, with a 3.6 cm diameter, D_o, circular opening located 20.7 cm above its base. The flow distribution was not measured, but such burners are designed to deliver uniform flows. The burner was located at one of three locations with the following centered flow exit positions: on the floor at the center of the enclosure (0.75 m, 0.75 m, 0.207 m), on the floor at the center of the rear wall with the exit edge 3.0 cm from the wall (0.75 m, 1.45 m, 0.207 m), and raised at the center of the enclosure with the burner exit 2.5 cm below the ceiling (0.75 m, 0.75 m, 0.725 m). Coordinates are given relative to an origin located on the floor at the front of the enclosure on the left-hand edge. Laboratory and gas temperatures were maintained at 21 °C ± 1 °C.

The average helium flow velocities, U_o, at the flow exit were 0.244 m/s and 0.058 m/s for the one hour and four hour releases, respectively. Following Chen and Rodi [67], for the higher flow velocity, the exit Reynolds number, Re, and Froude number, Fr, given by

$$\mathrm{Re} = U_o D_o / \nu_{He}, \tag{3}$$

$$\mathrm{Fr} = (\rho_{He} U_o^2)/(g D_o (\rho_{air} - \rho_{He})), \tag{4}$$

where ν is the kinematic viscosity and g the gravitational constant, are Re = 75 and Fr = 0.0271. The Reynolds number indicates the initial flow is laminar and, based on criteria given by Chen and Rodi [67], becomes buoyancy dominated approximately 2 cm above the flow exit. The four hour velocity flow will become buoyancy dominated even closer to the exit.

---

[1] Certain commercial equipment, instruments, or materials are identified in this paper in order to adequately specify the experimental procedure. Such identification does not imply recommendation or endorsement by the National Institute of Standards and Technology, nor does it imply that the materials or equipment are necessarily the best available for the purpose.

2.2.   Helium Sensor Calibration and Concentration Measurement

Measurements recorded during an experiment included helium volume fractions at eight locations. Two types of helium sensors were tested for these measurements: Neodym Technologies Panterra Model PN-ST-GHY-A040A-W20A-O5-R1-S1-E2-X0-I1-P2-L5-J1-Z0 and Xensor Integration Model TCG-3880. Both models respond to variations in the gas thermal conductivity as the helium concentration changes.

The Neodym Technologies sensors were optimized for hydrogen detection, but were calibrated for response to helium as described below. Each system consisted of a 2.5 cm × 2.5 cm × 2.5 cm sensor head connected by a 5 m cable to a 6.4 cm × 5.1 cm × 2.5 cm control box. The actual sensor had a circular surface of ≈ 1.3 cm diameter. The control box incorporated a digital microcontroller programmed to linearize the response of the thermal sensor and to correct for temperature sensitivity. A digital-to-analog converter generated a 0 V to 5 V output with roughly 1000 steps corresponding to the full range of the detector. The system was powered by a 5 V input. The quoted response time for the sensor system was 10 s.

Each sensor was individually calibrated in-house using a computer-controlled mixing system. [68] The mixing system was controlled by a legacy 486 personal computer running Strawberry Tree Workbench PC software to interface with a Strawberry Tree Flash-12 Model 1 data acquisition and ACAO-12-2 digital output board to control and read a 30 SLM (standard liters per minute) MFC for the air flow and two 10 SLM meters used in tandem for helium. Algorithms programmed in Workbench PC calculated the control voltages necessary for each MFC to provide flows of helium and air that, when mixed in a baffled chamber, provided a mixture with a preset helium volume fraction. The total volume flow rate was held constant at 29.9 L/min. The system was programmed to automatically vary the helium concentration over 21 separate steps of increasing helium volume fraction. The starting helium volume fraction as well as the step size could be modified by program inputs. Separate programs were used in which the volume fraction changed was stepped every 30 s or 60 s.

It proved difficult to program the mixing system to provide helium mixtures with well characterized and specified volume fractions. As an alternative, the delivered volume fractions were measured frequently with 1 % uncertainty by passing a portion of the flow through a pair of Siemens Calomat 6 7MB2527 gas analyzers. These gas analyzers have time constants around ten seconds. In order to provide sufficient time for the analyzers to fully respond, a period of 60 s between volume fraction steps for the mixing system was used. Tests with longer periods indicated a 60 s period was sufficient. Volume fractions measured by the analyzers were recorded for each helium concentration level at the end of each 60 s flow period. The helium volume fractions reported by the two analyzers agreed to within 0.001 helium volume fraction when the analyzers were zeroed and spanned with air and helium, respectively.

The sensors were calibrated by placing them inside a calibration cell formed from a 22.9 cm long PMMA tube with an inside diameter of 5.1 cm. The tube was o-ring sealed at either end with aluminum faces held in place by four threaded rods. One face contained a Swagelok fitting for introducing the mixture flow, while the other was open. The sensor to be tested was placed near the inlet end of the calibration cell, and the opposite end was sealed with a robber stopper with a 1 cm hole in the center to allow flow through the cell and the electric lead for the sensor to be passed into the cell. The calibration mixture volume flow rate was sufficient to sweep out the 0.47 L volume of the calibration cell in roughly 1 s. Testing showed that there was very little air backflow into the cell when the calibration mixture was flowing, even though air quickly entered when no flow was present.

During calibration, a sensor was exposed to the varying helium concentrations in the cell. The 0 V to 5 V output for the sensor was conditioned utilizing a SCXI 32-channel National Instruments 1102 signal conditioning board coupled with a SCXI-1300 interface board. The conditioned signal was fed to a PXI-6221 digitizer board. The digitizer board was controlled and read by a second personal computer running National Instruments LabVIEW software. Voltages were digitized at a 2 kHz rate and then averaged for 1 s before being saved in a comma-delimited file along with the relative time of the sample. Typically, two sets of calibrations would be run for each sensor, one recorded for helium volume fractions

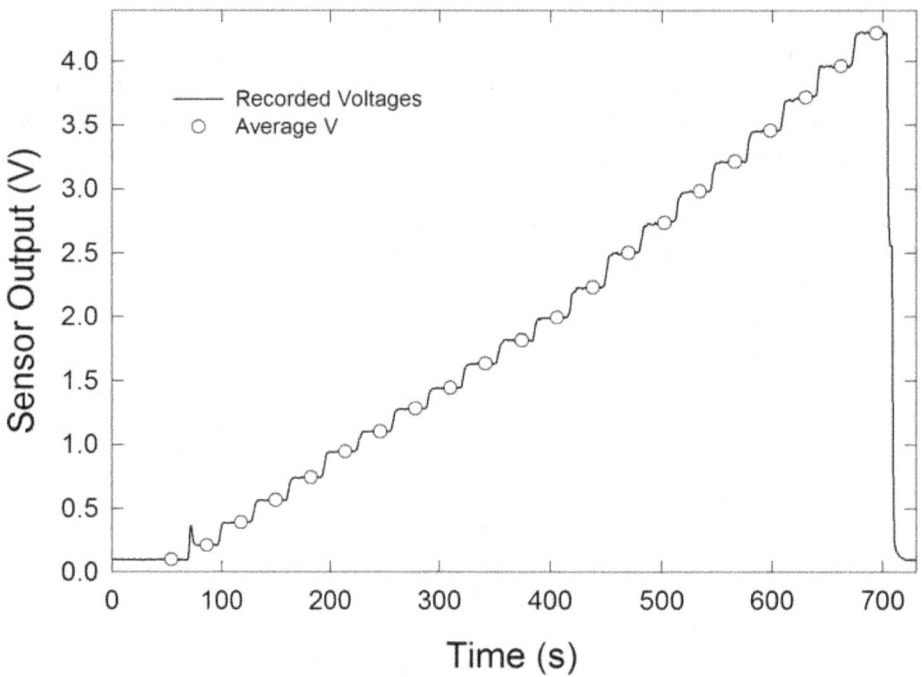

Figure 2.   The measured output voltage for a Neodym Technologies thermal conductivity sensor is shown as a function of time as the helium volume fraction is stepped over a 0 to 1 range.

varying from 0 to 1 in nominal steps of 0.05 and the second from 0 to 0.2 in nominal steps of 0.01. Figure 2 shows an example of one of these calibrations plotted as sensor output voltage versus time. It is evident that the sensor responds fully to the changes in the helium concentration in less than 5 s even though this period is less than the stated response time.

Results for the calibrations were plotted utilizing SigmaPlot software. A SigmaPlot transform was created to average the voltage outputs over a 20 s period running from roughly 8 s to 28 s of a given volume fraction step. The circles included in Figure 2 indicate these averages for the 21 volume fraction steps. Individual calibration curves for each sensor were generated by fitting fourth order polynomials to plots of delivered helium volume fraction versus sensor output voltage for the two calibration ranges. Figure 3 shows an example of one of these fits. Using the fourth order polynomial it was possible to calculate the helium volume fractions for an arbitrary voltage output from this sensor.

During the calibrations of the Neodym Technologies sensors small sharp discontinuities in voltage were identified at several concentrations over the calibration curve. These jumps increased the uncertainty in the measurements around these volume fractions. The reason for these jumps was not identified, but it seems likely that they, as well as the nonlinearity in the calibration curve evident in Figure 3, were associated with sensor software corrections made to the response assuming hydrogen was being measured, while helium was actually being measured.

The Xensor Integration sensors, which also respond to variations in gas thermal conductivity with helium concentration, function in a different way than the Neodym Technologies sensors. These sensors are manufactured using solid-state techniques. The sensor element is a small chip of silicon with rectangular dimensions of 2.5 mm × 3.3 mm and a thickness of 0.3 mm. A small isolated silicon nitride membrane is fabricated in the silicon with a small heater located in the center. The application of a small electric current raises the temperature of the heater above the ambient temperature of the surrounding silicon. Heat is lost from the center by thermal conductivity through the silicon nitride membrane to the silicon chip, which is held at the ambient temperature and to the surrounding gas, with the latter depending on the gas thermal conductivity. A thermopile formed from 6 n-doped and 6 p-doped

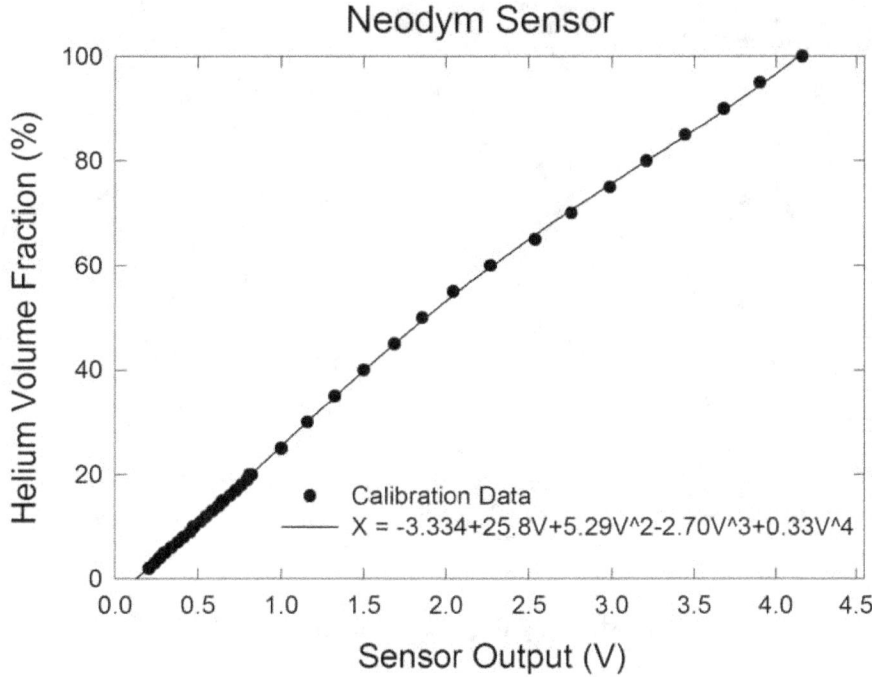

Figure 3.   An example of the fourth order polynomial fit to experimental data for the response of a Neodym Technologies sensor to varying helium volume fractions.

polysilicon strips senses the temperature difference between the heated center and the edge of the silicon nitride membrane. The thermopile voltage output is the sensor signal.

The manufacturer indicates these sensors can be used to measure absolute low pressures, material properties for very small samples, and the composition of binary gas mixtures.   They have been widely used as microcalorimeters to measure thermal properties of small samples of material subjected to high heating rates. Of more direct interest is their use to measure helium and hydrogen concentrations. Gupta et al. used these sensors to measure concentrations when helium was released into a full-scale garage. [14] The sensors were individually calibrated by measuring their response to known helium/air mixtures covering a 0 % to 50 % range. A second-order polynomial was then fit to a plot of sensor output voltage versus concentration. Denisenko et al. reported similar measurements using these sensors for hydrogen released into a sealed vessel. [69] The sensors were calibrated in hydrogen/air mixtures generated by a specially designed mixing system that spanned the 0 % to 100 % range. Plots of sensor response versus hydrogen concentration were also fit to second-order polynomials.

The TCG-3880 sensors were delivered mounted on a standard TO-5 ten-pin mount, with a circular base having a 9.4 mm diameter. Due to their operating principle, the sensors are sensitive to changes in ambient temperature as well as flows over their surfaces that increase the heat loss. As an option, Xensor Integration will include a Pt100 class B platinum resistance thermometer mounted along side the TCG-3880 sensor that can be used to measure temperature in order to allow correction for ambient temperature variations. Both TCG-3880 and TCG-3880Pt versions were tested. Since the ambient temperature in the laboratory was nearly constant, no attempts were made to read and correct for temperature variations. The manufacturer indicates that the temperature sensitivity can be minimized by controlling the current flow to the sensor heater by including an external 2 kΩ resistor with a small thermal coefficient in series with a voltage source. This recommendation was followed as was that to power the heater with a 4 V input.

The sensors were inserted into a TO-5 socket supplied by Xensor Integration mounted on a small PCB board with 10 holes for wire connections.   The PCB board was attached to a small section of perforated board that was used to hold the resistor for the heater circuit. This entire assembly was

Figure 4.   Photograph of a TCG-3880 showing the associated mount, external heater resistor, and wiring.

mounted on a small machined aluminium mount that was used to support the sensor at a given location. Figure 4 shows a close-up photograph of one of the sensors as it appeared in use.

As configured, the sensors are expected to generate voltage outputs of roughly 45 mV and 10 mV when placed in air and helium, respectively. The sensor output voltages were digitized using the same data acquisition system described above for the Neodym Technologies sensors even though the voltage ranges differed by a factor of roughly 100. As discussed further below, the voltage resolution of the system was adequate when using the TCG-3880 sensors.

Preliminary tests in the calibration cell indicated that bare TCG-3880 sensors were sensitive to the flow inside the cell. In order to minimize this response, the sensors were sealed inside a standard TO-5 cover having a 5 mm height and 8 mm diameter. A 0.28 cm diameter circular hole was drilled in the center of the cover top to facilitate gas exchange between the inside and surroundings. The cover limited gas flow over the sensor and greatly reduced the sensitivity to flow. Some sensitivity was still evident when a covered sensor was placed directly into the flow entering the calibration cell with its cover opening facing the flow. However, no flow sensitivity was evident with the sensor placed outside of the inlet jet and oriented perpendicular to the primary flow direction. As will be shown shortly, the time response of the covered sensor to a change in concentration was satisfactory for the current application.

Figure 5 shows an example of the sensor output voltage versus time for a TCG-3880 sensor in the calibration cell with the nominal volume fraction of helium varying over a range of 0 % to 20 % in steps of 1 %. The open symbols represent average values for a given condition. The mixing system had trouble controlling the flows for the three lowest helium concentrations, and average values are not included for these casess. While some noise is evident on the signals, the noise levels are small compared to the step sizes corresponding to 1 % changes in helium volume fraction. As a rough indication of the noise level, a typical voltage step for the data in Figure 5 is 0.5 mV, while the corresponding root mean square (rms) values for the averaged measurements have values between 0.02 mV and 0.03 mV. These results indicate that it should be possible to resolve helium concentration changes on the order of 0.1 %.

Calibration curves for the Xensor Integration sensors were generated in the same way as for the Neodym Technologies sensors, i.e., by fitting fourth-order polynomials to data from calibrations over 0 % to 20 % and 0 % to 100 % ranges of helium volume fraction. An example of such a fit is shown in Figure 6. The result of the fit is included on the plot as a solid curve. The agreement between the calculated curve and the experimental data is very good, as indicated by the correlation coefficient of 0.9999985 for

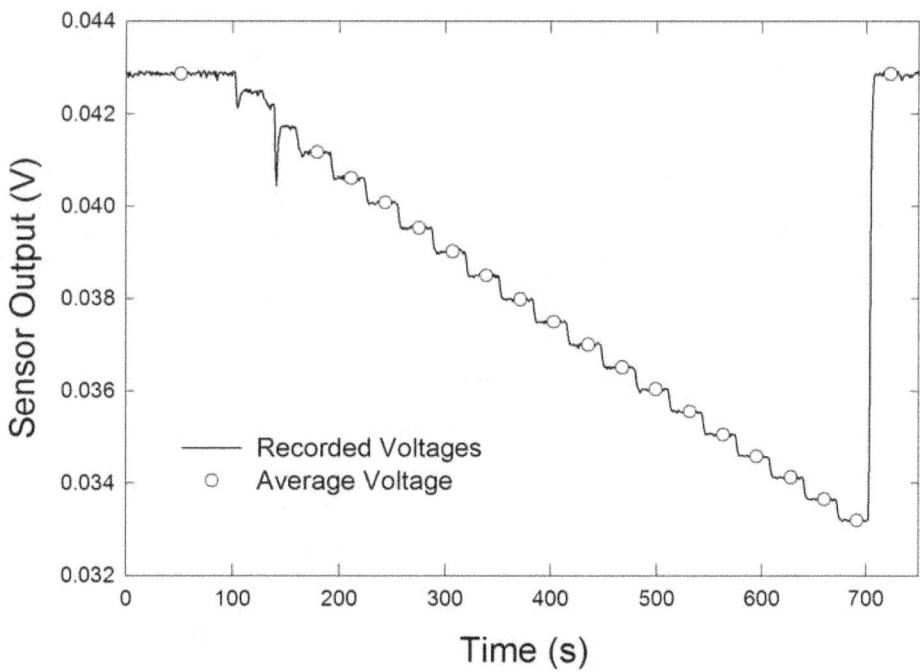

Figure 5.   The measured output voltage for one of the Xensor Integration thermal conductivity sensors is shown as a function of time as the helium volume fraction is stepped over a 0 to 0.2 range.

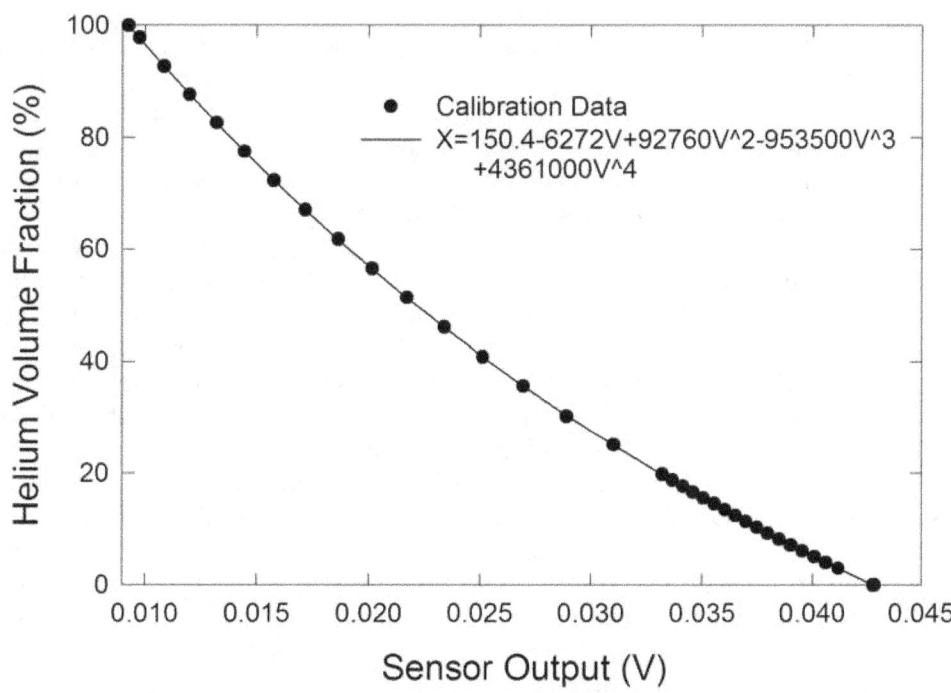

Figure 6.   An example of a fourth order polynomial fit to experimental data for the response of a Xensor Integration sensor to varying helium volume fractions.

the fit. Such high correlation coefficients were typical for these sensors. The results of the fit can be used to calculate the helium volume fraction for any arbitrary measured sensor voltage.

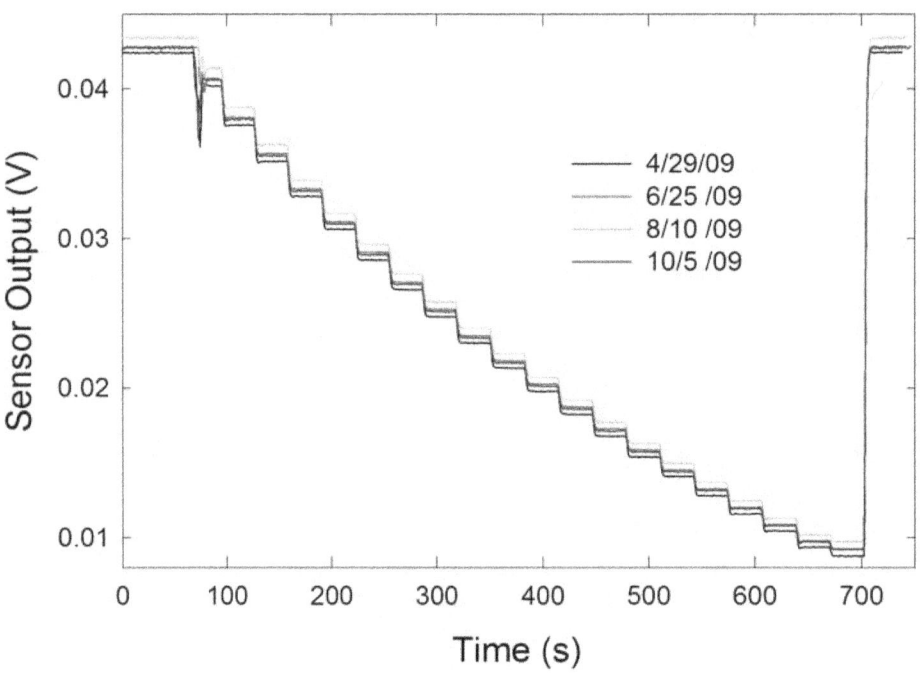

Figure 7.   Four repeated calibrations over a five-month period for the response of a Xensor Integration probe to helium volume fractions ranging from 0 to 1 are shown.

Figure 7 shows four repeated calibrations for one of the TCG-3880 probes made over a five-month period. These calibrations covered the full range of helium volume fraction in nominal steps of 0.05. Some variations in output voltage are evident. Close inspection revealed that these variations resulted from baseline shifts, and that the relative voltage values across the calibration curve were unchanged. No obvious reason for the baseline shifts was identified. Measurements of sensor response lasting several days suggested that the changes occurred relatively slowly.

The behavior of the baseline shifts evident in Figure 7 suggested an approach for correcting for them. When the response of a given sensor to air is known, it is possible to calculate the expected voltage for the sensor using the fourth-order calibration coefficients for that sensor. If there is a difference between the measured and predicted values, the difference can be subtracted from the experimental value so that the voltage is shifted and zero helium concentration is calculated. In principle, this correction will apply over the entire helium concentration range. In practice, this was achieved by recording measurements in air, see below, and adjusting the experimental voltage readings as described. Comparisons of experimental data covering a range of helium volume fractions indicated this approach was successful, with helium volume fractions for a number of sensors calculated using repeated individual calibration curves yielding very similar volume fractions with no differences larger than 0.003.

Some additional tests used to characterize the response of the Xensor Integration sensors are described here. Figure 8 shows a calibration curve for one of the sensors on expanded time and voltage scales. The voltage drop due to one of the 0.05 step decreases in helium volume fraction is clear. Roughly 3 s is required for the voltage output to fully attain the new level, with much of the change occurring during the first one-second time step. This observation means that the sensor response time must be on the order of 1 s or less. This is particularly true when it is recognized that the observed voltage response time is also convoluted with the one-second averaging time used for the digitization, the response time for the mixing system, and the mixing time within the calibration cell.

12

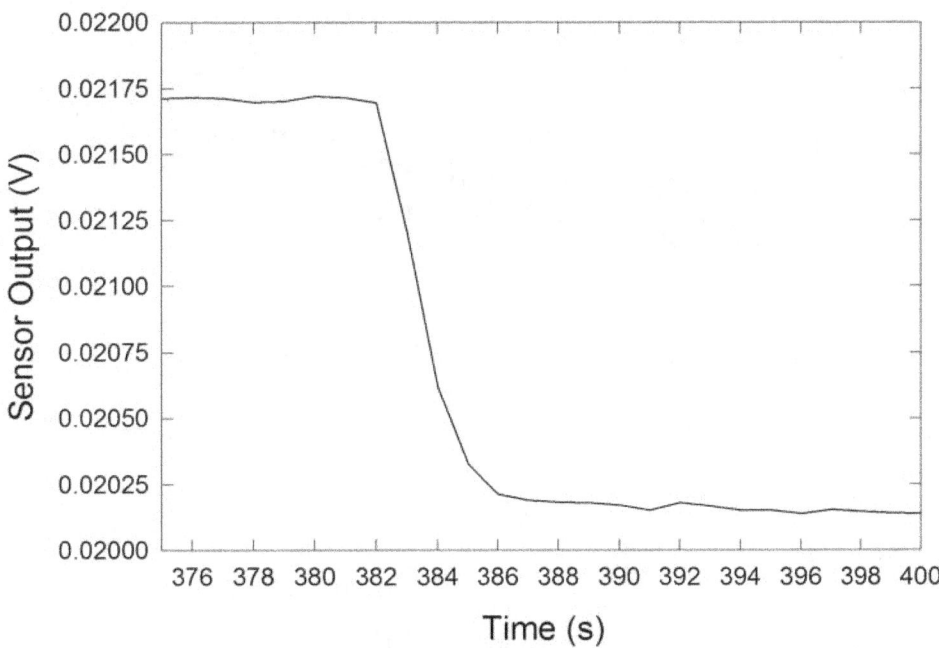

Figure 8. A portion of the data shown in Figure 5 is blown up in order to emphasize the sensor time response.

By placing tape over the vent(s) in the reduced-scale garage, it was possible to seal the enclosure following a helium release. After a period of time molecular diffusion smoothed out the vertical helium volume fraction gradients, and the concentration became uniform. Since any leaks from the garage were very small, the helium concentration was essentially constant with time. Figure 9 shows helium volume percents measured by eight individually calibrated Xensor Integration sensors for a test in which the enclosure was sealed 0.45 hr after the end of a one hour release of helium. The helium release was started at 60 s. Systematic differences in concentration readings covering a helium volume fraction percent range of 0.4 % are evident for the different sensors. These differences are likely due to randomness in the calibration process and thus provide an indication of type A uncertainty. It can also be seen that the concentration readings for individual sensors varied slightly with time, with the helium volume fraction increasing by $\approx$ 0.2 % over a 2½ day period. Two possible explanations for these small increases are slow baseline drifts in the calibration curves as discussed earlier or slow diffusion of helium into the enclosure from the line that fed the helium flow system. In either case, the plot shows that very little drift in measured volume fraction readings is to be expected over periods on the order of a day.

A similar experiment was performed using the front wall with vents near the floor and ceiling. A one hour release of helium into the enclosure was carried out with the top vent sealed in order to build up a high helium concentration inside the enclosure. Shortly following the end of the release, the lower vent was also sealed, and the helium concentration gradients inside the enclosure were allowed to dissipate. After a period of time both, vents were unsealed, and the resulting exchange with the outside caused the overall interior helium concentration to drop as concentration gradients again developed. After some time the vents were again sealed, and the interior developed a uniform, but lower, concentration than before. This process was repeated several times to allow the sensors' response to a range of concentrations to be characterized.

Figure 10 shows plots of helium volume fractions versus time for the eight sensors. The periods when the vents were open and closed are easily identified. A portion of the plot is shown blown up in Figure 11. Concentration gradients began to develop as soon as the vents were opened at 18.32 hr, and the helium volume fractions began to drop. When the vent was resealed after 0.38 hr the concentration

13

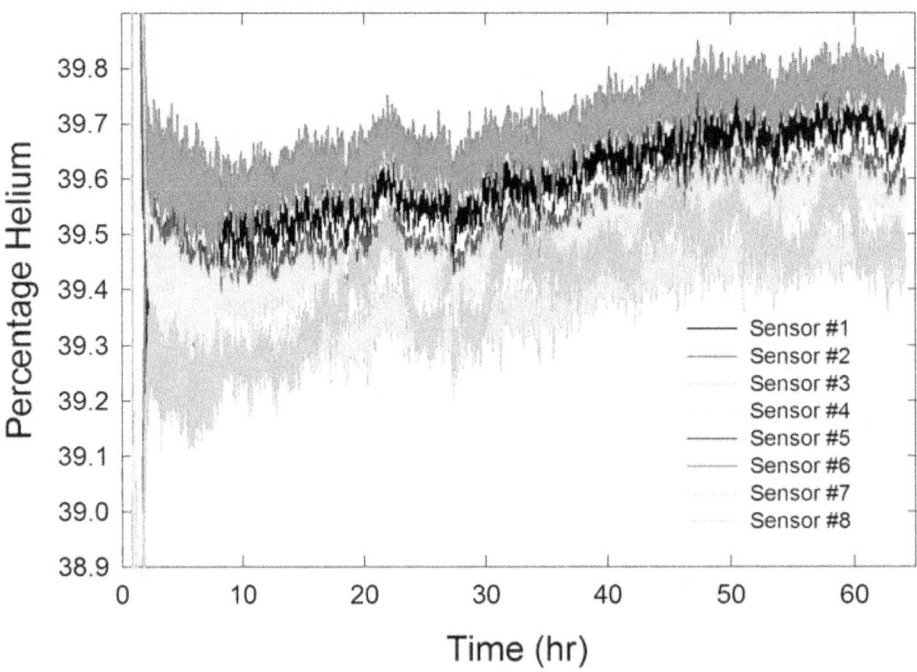

Figure 9.   The percentage of helium recorded by Xensor Integration sensors over a roughly two and a half day period at eight locations within a sealed enclosure are shown.

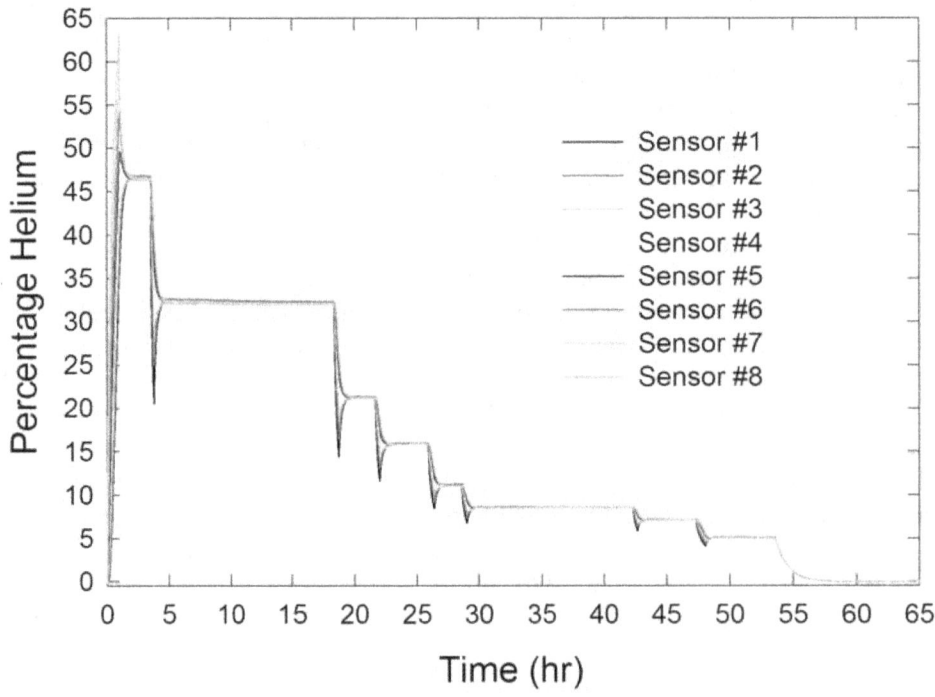

Figure 10. Helium volume fractions measured with Xensor Integration sensors are plotted as a function of time for eight locations within the enclosure which was sealed and unsealed multiple times. A flow of air was introduced into the enclosure at around 53 hours.

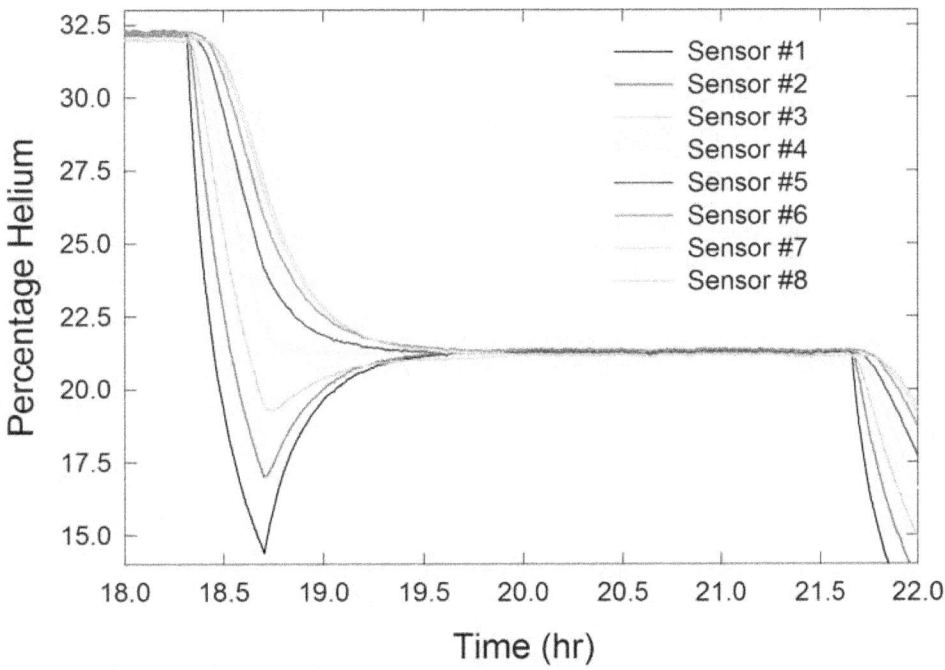

Figure 11. The data shown in Figure 10 is replotted on expanded percentage helium and time scales.

gradients immediately began to dissipate, with concentrations in the bottom of the enclosure rising and those near the top falling. In less than an hour the concentrations were again nearly uniform.

Average helium volume fractions were calculated for each sensor during periods when the concentrations were constant by averaging over 2000 s periods. Figure 12 shows the averaged value for each sensor plotted against the average for all of the sensors taken together. The results for different sensors lie very close together. Root mean square (rms) values obtained when averaging the helium volume fraction results for the eight sensors ranged from 0.13 % for the highest concentrations down to 0.04 % for the lowest. These values confirm the excellent agreement for measurements recorded by the independently calibrated sensors over this concentration range. The rms values provide quantitative estimates for the Type A uncertainty due to the calibrations.

While it appears that uncertainties in helium volume fraction associated with variations between repeated calibrations for different sensors were much less than 0.005, the absolute uncertainties in the measurements were likely somewhat larger since the uncertainty in helium volume fraction measurements made by the Siemens thermal analyzers were not fully characterized beyond the manufacturer's specification of $\pm$ 1 % of the span range.

If the thermal conductivity dependence on helium concentration in helium-air mixtures is available with sufficient accuracy, it should be possible to predict the response of the Xensor Integration sensors. As part of a project to be described elsewhere in which these sensors were calibrated for use in hydrogen/air mixtures, a model was developed that utilized sensor voltages measured in pure air and helium along with thermal conductivities for the pure gases and mixtures to predict the sensor response for arbitrary mixtures. The starting point for the analysis was the sensor response which, according to the manufacturer, can be written as

$$V_{out} = P_{in} \times S_{tp} / (G_{mem} + G_{gas}), \tag{5}$$

where V_{out} is the output voltage for the sensor, P_{in} is the electrical input to the sensor heater, S_{tp} is the thermopile sensitivity, and G_{mem} and G_{gas} are the thermal conductances through the sensor membrane and

15

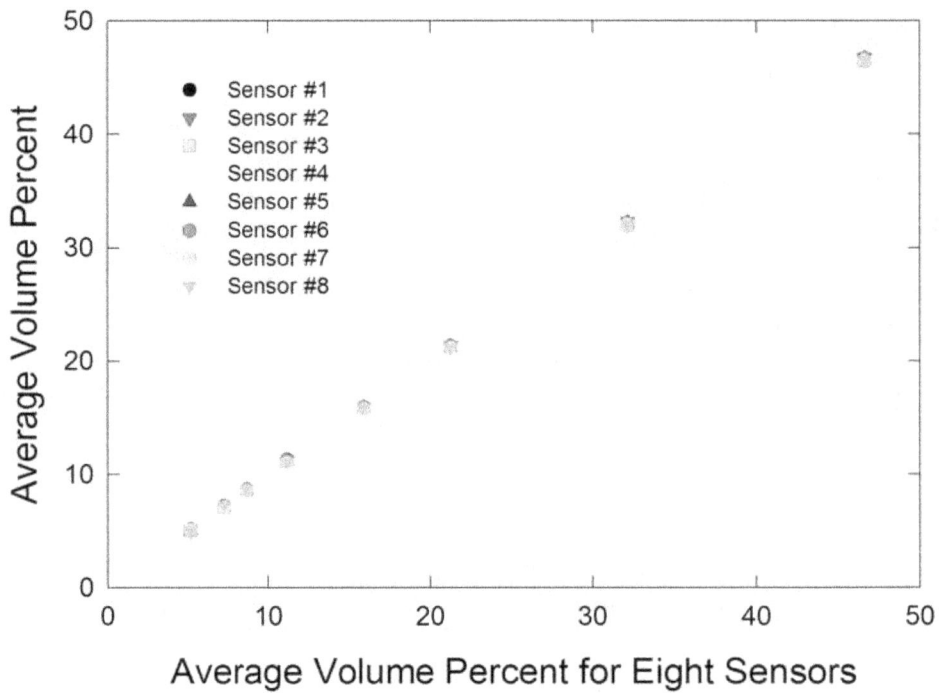

Figure 12. Average helium volume percents measured by eight Xensor Integration sensors are plotted against the corresponding average for all of the sensors for a range of helium concentration in the reduced-scale garage.

gas, respectively. For the operating conditions used here P_{in}, S_{tp}, and G_{mem} are very nearly constant, and Eq. (5) can be rewritten as

$$V_1 = c_1 /(c_2 + k_1),$$

(6)

where c_1 and c_2 are constant, and k_1 is the thermal conductivity for a given gas. For a second gas the expression can becomes

$$V_2 = c_1 /(c_2 + k_2).$$

(7)

Equations (6) and (7) can be solved simultaneously to give

$$c_2 = \frac{V_2 k_2 - V_1 k_1}{V_1 - V_2}.$$

(8)

and

$$c_1 = V_1 (c_2 + k_1).$$

(9)

By measuring the response of a sensor to two gases with known thermal conductivities, values of c_1 and c_2 for the sensor could be determined and then used in Eq. (5) to predict the sensor response for an arbitrary gas having a known value of thermal conductivity. This approach was used to predict the response for one of the Xensor Integration sensors using experimental values for the sensor outputs in air

16

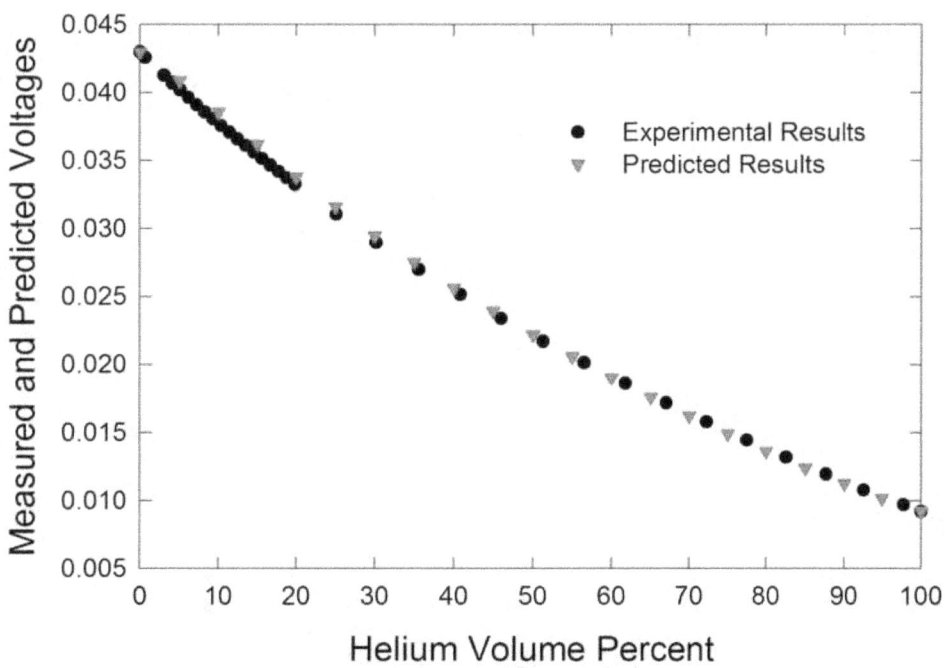

Figure 13. Experimental and predicted voltages for the response of a Xensor Integration sensor to a range of helium concentrations is shown.

and helium along with recommended values of thermal conductivity for helium/nitrogen mixtures at 303.2 K taken from the TPRC data series. [70] It was considered appropriate to use helium/nitrogen mixture results since air and nitrogen have very similar thermal conductivities.

The TPRC recommended values are available in helium volume fraction percentage steps of 5 % over a range from pure air to pure helium, and predicted sensor voltages were calculated over this range. The predicted response is compared with experimental measurements for the same sensor in Figure 13. Visual inspection shows that the two curves fall close to each other. In order to quantify the agreement, the two curves were fit to fourth order polynomials, and the predicted curve was subtracted from the fit for the experimental results. A plot of the difference as a function of helium volume percent is shown in Figure 14. The two curves vary systematically over the full concentration range, with maximum absolute differences in volume fraction on the order of 1.5 %. The observed differences are comparable to the absolute uncertainty of 1 % helium volume percent specified by Siemens for their thermal conductivity meters. Since measurements by individual sensors provide measurements with less noise than this, and measurements by multiple sensors yielded values that agreed to within less than 0.5 %, an absolute value of 1 % was adopted as the largest uncertainty for measurements of helium volume fraction recorded using Xensor Integration sensors.

2.3.   Pressure Measurement

A single pressure port was located on the front face at a location 37.5 cm from the left wall and 37.5 cm above the floor. The differential pressure between the enclosure interior and ambient surroundings was measured with a 133 Pa (1 torr) Baratron electronic manometer. The manometer was calibrated by determining its voltage output versus a range of pressures generated by rotating the shaft of an Otis Model MB 21 HT bellows pump and measured with a Dwyer Model 1430 Microtector Portable Electronic Point Gage. The quoted uncertainty for the pressure measurement is 0.06 Pa. Figure 15 shows a plot of the differential pressure (ΔP) in torr versus the measured Baratron voltage for measurements recorded on three different days. The straight line is the result of a linear least squares curve fit to the

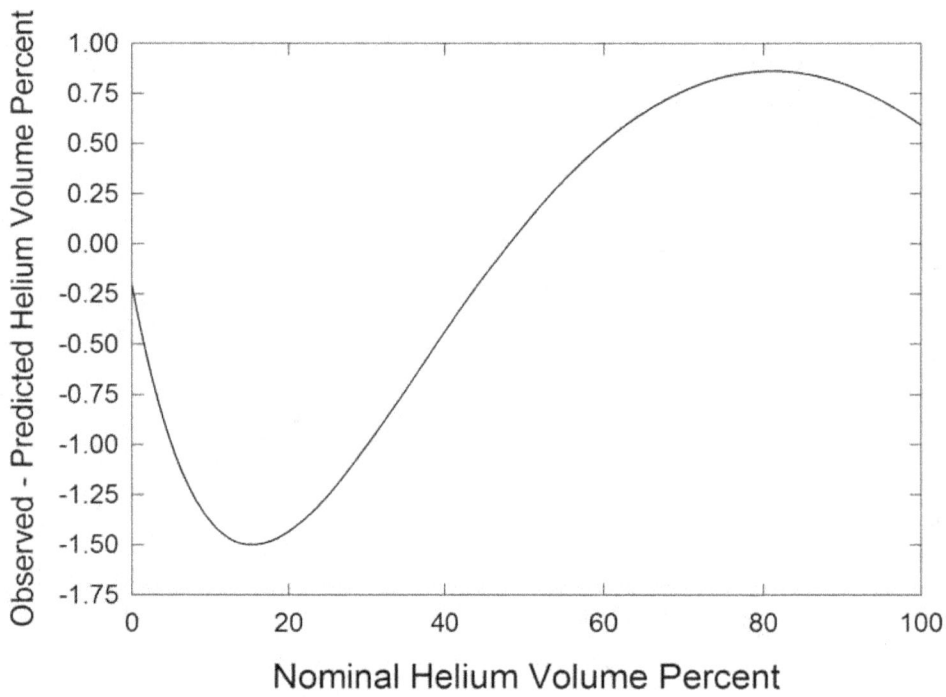

Figure 14. The difference between experimental values of helium volume fraction and values calculated using the approach described in the text plotted as a function of the nominal helium volume percent.

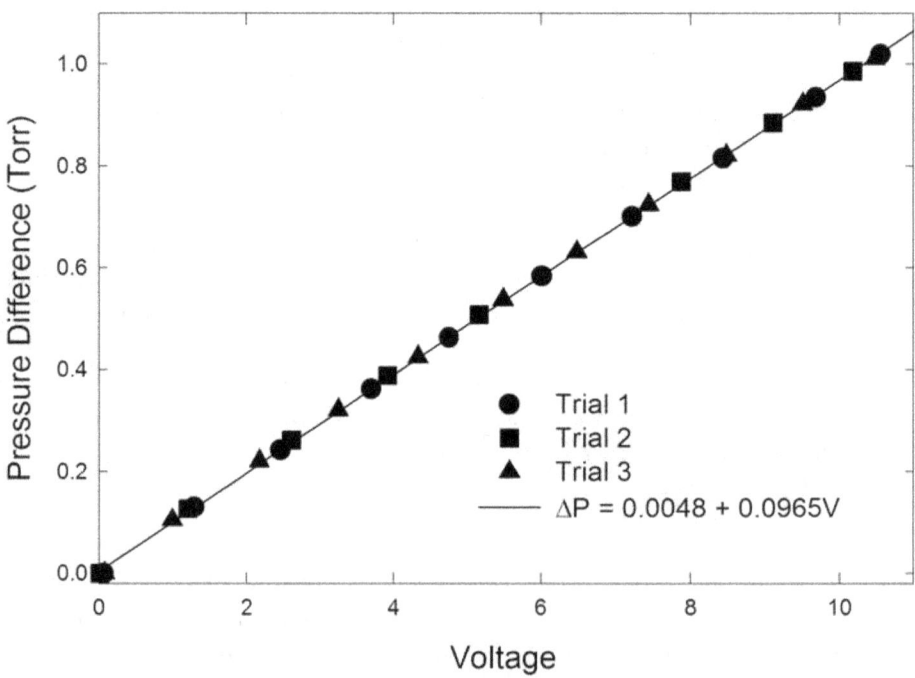

Figure 15. The results for three calibrations of the Baratron electronic manometer are plotted as differential pressure versus voltage. The straight line is a linear least squares curve fit to the data.

Table 1. Heights for Vertical Array of Seven Helium Sensors Located 37.5 cm from the Front and Side Walls of the Reduced-Scale Garage

Sensor Number	Height above floor (cm)
1	9.3
2	18.5
3	27.6
4	37.2
5	46.6
6	55.9
7	65.0

data. The calculated slope and intercept for the data were 0.0965 ± 0.0002 and 0.0048 ± 0.0015, respectively. The uncertainties are very low, indicating an excellent fit.

2.4. "Doorway Fan Test"

A doorway fan test, also commonly referred to as a blower doorway test, is used to characterize leaks into or from an enclosure. [62,63] A fan is sealed into a doorway or window in the enclosure to be tested, and the pressure drop across the opening is measured as the fan speed and calibrated air volume flow rate is varied. Analogous tests were performed for the enclosure studied here by passing known volume flow rates of air through the Fischer burner and measuring the resulting pressure drop across the enclosure interface. A combination of mass flow controllers and dry test meters was used to measure air volume flow rates. Values of C and n were obtained by fitting the measurements to Eq. (2) using a least squares curve fit procedure.

2.5. Time-Resolved Helium Volume Fraction Measurements in the Reduced-Scale Garage

Both Neodym Technologies and Xensor Integration sensors were used to record time-resolved helium volume fraction measurements inside the enclosure described below. In general, the two types of sensors provided measurements that agreed quite well. Due to their ease of use and somewhat smaller sensor size, we chose to use Xensor Integration sensors for all of the helium volume fractions measurements reported here.

A group of seven Xensor Integration sensors was assembled along a vertical line located 37.5 cm from the left and front walls. Table 1 lists the assigned numbers and heights for these sensors. An eighth sensor was moved to variable positions during different tests to check the horizontal uniformity of the helium distribution within the enclosure.

The voltage outputs for the sensors were digitized by simultaneously recording the signals on separate input channels of the data acquisition system described earlier. The LabVIEW program that controlled the data acquisition was written so that the data collection rate could be changed during an experiment at a preset time. The full-range 0 V to 5 V output of the Baratron pressure transducer was digitized at the same time and rate to provide a measure of the differential pressure between the inside and outside of the enclosure at the transducer height.

A typical helium release experiment consisted of recording a short background level in air (typically 60 s) before initiating the helium flow, starting the helium flow appropriate for a one hour or four hour release, halting the helium flow at the appropriate time, and recording the pressure and concentration sensor voltages for variable periods (sometimes up to several days) during the post release period. Measurements during the release phase were always recorded with a 1 s averaging time, while it was common to switch to the 10 s average at some point during the post-release period in order to limit the number of samples recorded.

It was typical to either allow sufficient time for all of the helium to exit the enclosure or to initiate a known air flow (29.6 L/min $\pm$ 0.1 L/min) generated by a mass flow controller to sweep out any remaining helium. When all of the helium was removed, the air flow was halted, and the pressure transducer voltage was recorded for the no flow condition. By averaging over the periods just prior to and after the time when the air was halted, the pressure increase due to the known air flow was obtained by difference. These measurements are similar to the fan tests for the different vent configurations and were used to assess reproducibility and to check for changes in the system over time.

The voltage readings for the eight thermal conductivity sensors and the pressure transducer were saved continuously in a named comma-delimited file. Conversion of the measured voltages to volume fractions and pressures along with data plotting was accomplished by importing the voltage signals into SigmaPlot and using a specially written data transform to implement the individual calibration curves. The transform for the concentration measurements corrected for any baseline drifts in the Xensor Integration sensors by first calculating the voltage outputs for each of the sensors in air by averaging values over the initial 10 s to 50 s of the time record, which were recorded prior to initiating the helium flow. The differences between these values and the corresponding voltages for 100 % air determined from the 4th order polynomial fits to the calibration data were subtracted from the voltage time records for each sensor. The helium volume fractions were then calculated using the offset voltages along with the individual calibration coefficients. It was observed that the pressure transducer produced a small negative voltage for zero differential pressure. The magnitude of this signal was determined by averaging the output voltage for the pressure transducer over the same 10 s to 50 s period used for the concentration signals and then using this value as the intercept for the calibrated linear response.

Once the time records were available for the eight sets of volume fraction measurements and the differential pressure, the results were plotted using SigmaPlot.

3.   Experimental Results

3.1.   Fan Test Characterization of Vent Configurations

Figure 16 shows the results of three repeated fan tests for the enclosure equipped with a front wall having a single 2.4 cm × 2.4 cm square opening in the center plotted as air volume flow rate versus the measured differential pressure induced by the flow. The data was fit to Eq. (2) using a non-linear least squares curve fit procedure. The result of the fit is shown in the figure as a solid line.

The exponent for the differential pressure for a given orifice is known to depend on the type of flow through the opening, with values varying from 0.5 for purely turbulent flow to 1 for purely laminar flow. The experimental exponent of 0.531 indicates the flow through the single opening is turbulent. This is the expected behavior for this size opening and volume flow rates.

Using the coefficients derived from the fit along with Eq. 2 gives the volume flow rate for a 4 Pa differential pressure of $(Q_{enc})_{4 Pa}$ = 0.000929 m^3/s, which was the value used in Section 2.1. This indicates that the flow coefficient, C_v, necessary to relate the flow rate calculated using the Bernoulli equation (Eq. 1) to that actually measured is 0.66.

Figure 17 shows the corresponding results for the enclosure equipped with the front wall having a single 3.05 cm × 3.05 cm opening in the center. The pressure coefficient of 0.474 is consistent with the expected value of 0.5. The calculated flow rate for a differential pressure of 4 Pa is $(Q_{enc})_{4 Pa}$ = 0.00152 m^3/s. This corresponds to $(ACH)_{4 Pa}$ = 3.2, which is close to the target value of $(ACH)_{4 Pa}$ = 3.0.

Figure 18 shows the fan test results for the enclosure equipped with the front wall having two 2.15 cm × 2.15 cm vents. The calculated flow rate for a 4 Pa differential pressure is $(Q_{enc})_{4 Pa}$ = 0.00160 m^3/s corresponding to $(ACH)_{4 Pa}$ = 3.4. This result is similar to, but slightly larger than measured for the front wall with a single large opening. This is the expected result since the areas for the large vent and total of the two small vents are nominally the same. The small difference is likely due to uncertainties in the dimensions of the openings and in the parameters derived from the fan test measurements.

20

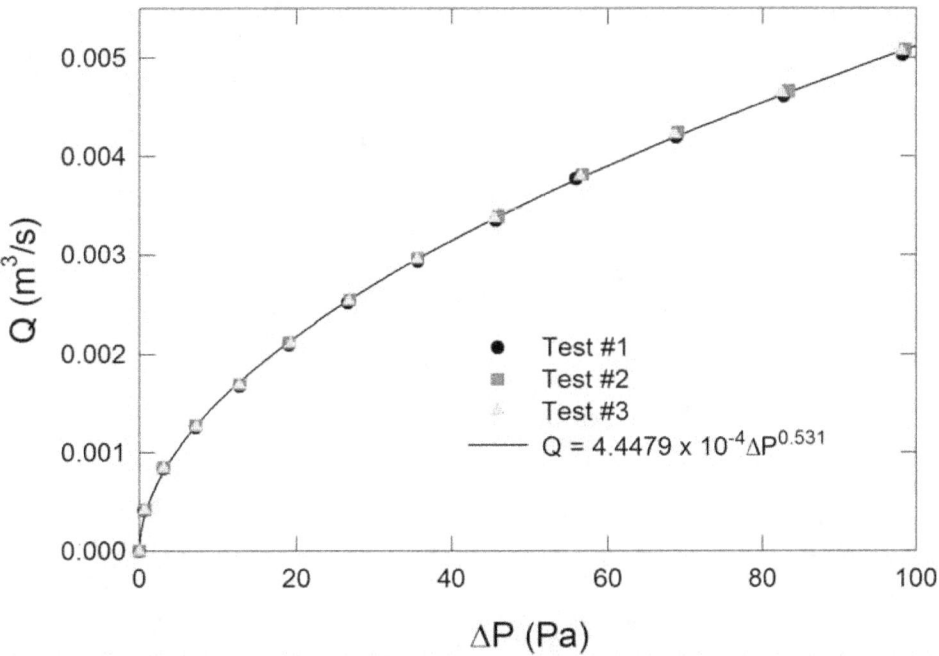

Figure 16. Values of air flow rate are plotted against the measured differential pressure for three repeated fan tests of the reduced-scale garage with a single 2.4 cm × 2.4 cm opening in the front wall. The solid line is the result of a non-linear least squares curve fit to Eq. (2).

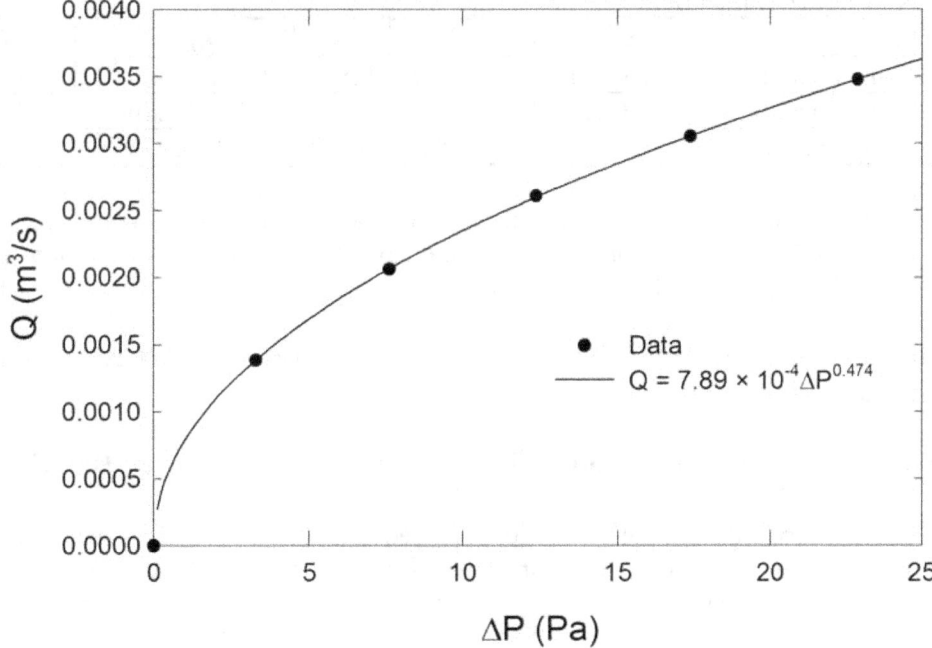

Figure 17. Values of air flow rate are plotted against the measured differential pressure for a fan tests for the reduced-scale garage with a single 3.05 cm × 3.05 cm opening in the front wall. The solid line is the result of a non-linear least squares curve fit to Eq. (2).

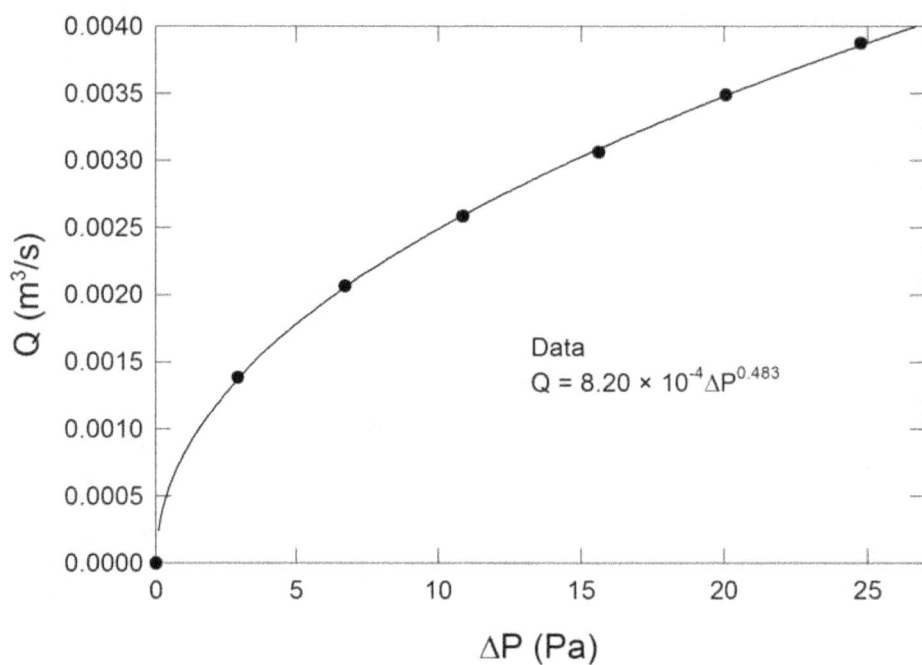

Figure 18. Values of air flow rate are plotted against the measured differential pressure for a fan tests for the reduced-scale garage with two 2.15 cm × 2.15 cm openings in the front wall. The solid line is the result of a non-linear least squares curve fit to Eq. (2).

Table 2. Front Wall Vent Characteristics

Vents	Area	C	n	$(Q_{enc})_{4\,Pa}$	$(ACH)_{4\,pa}$	C_v
2.4 cm × 2.4 cm opening	5.76 cm²	0.000445	0.531	0.0009286 m³/s	1.98	.660
3.05 cm × 3.05 cm opening	9.30 cm²	0.000789	0.474	0.001522 m³/s	3.24	.634
Dual 2.15 cm × 2.15 cm openings	9.25 cm²	0.000820	0.483	0.001602 m³/s	3.42	.671

The results of the fan tests for the three faces are summarized in Table 2. Parameters included are total vent area, values of C and n, the predicted volume flow rate for a 4 Pa differential pressure, the predicted value of $(ACH)_{4\,Pa}$, and the flow coefficient.

3.2.   Time-Resolved Helium Concentration and Pressure Measurements in the Reduced-Scale Garage

3.2.1.   One Hour Helium Releases near the Floor at the Center of the Garage

The parameters varied during the test series included helium flow rate (2), vent size and location (3), and helium release point (3), which yielded an experimental test matrix of eighteen experiments using all possible combinations. The results presented in this report were recorded over a period of six months. The sensors were calibrated at the beginning and end of the test period. The calibrations showed that the sensors, with the exception of the baseline shifts discussed earlier, were very stable. The calibrations recorded following the test series were used to calculate helium volume fraction for the tests.

A large number of similar experiments were completed prior to the current work using different groups of sensors. Some of the results from these earlier measurements were included in the paper presented during the Third International Symposium on Hydrogen Safety. [43] Comparison show only minor differences between the results presented here and those included in the earlier paper.

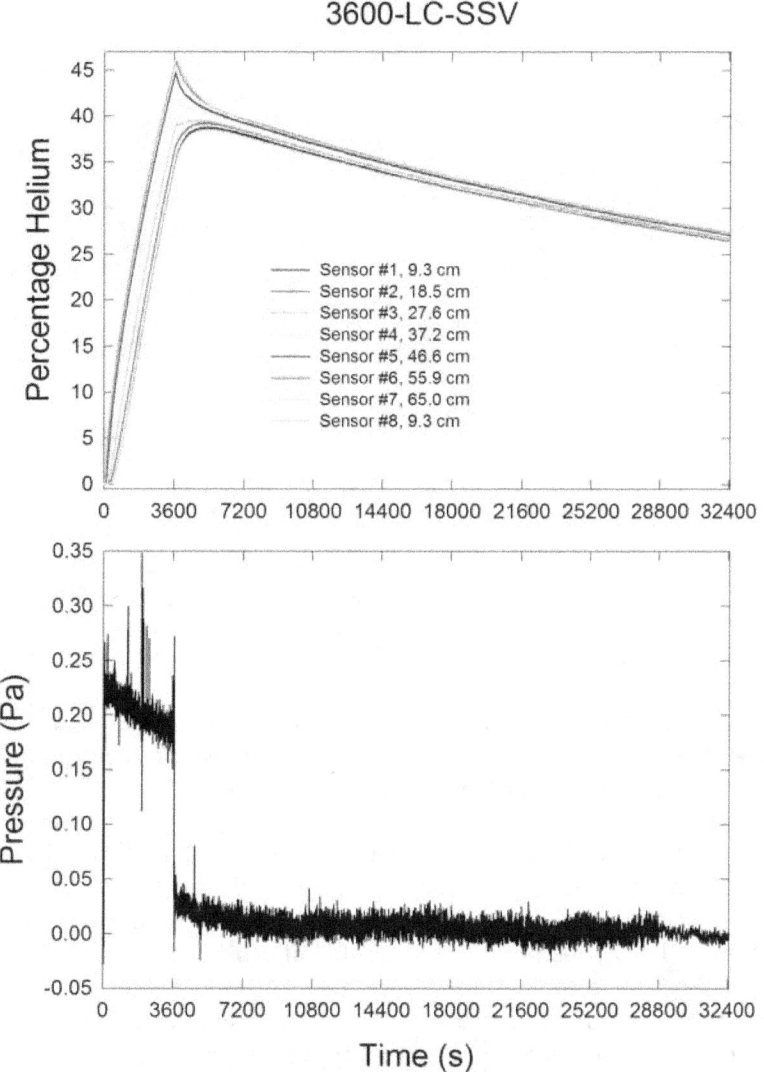

Figure 19. The upper plot shows helium volume percent as a function of time at eight locations for a one hour release of helium from the lower center position into the ¼-scale garage with a single centered 2.4 cm × 2.4 cm opening in the front face. The lower plot shows the differential pressure across the face.

A short-hand nomenclature is used to identify the different tests. A given test is assigned an identifier #####-xx-yyy, where ##### represents the helium release period in seconds and is either 3600 or 14400, xx denotes the burner location and has values of LC (lower center), UC (upper center) or LR (lower rear), and yyy is the vent configuration and takes values of SSV (single small vent), SLV (single large vent), or ULV (upper-lower vent).

The upper plot of Figure 19 shows measured helium volume percent for the eight sensor locations versus time for a one hour release of helium from the center position near the floor $((x,y,z) = (0.75 \text{ m}, 0.75 \text{ m}, 0.207 \text{ m}))$. The heights of the sensors are indicated. Note that the helium flow started at 60 s. The total time period shown is nine hours. In many cases data was collected for much longer periods, but nine hours was judged adequate to characterize the release and post-release portions of the concentration time profiles.

Sensors #1 to #7 were located along a vertical line located at $(x,y) = (0.375 \text{ m}, 0.375 \text{ m})$. For this experiment sensor #8 was placed at the same height as sensor #1, i.e., $z = 0.093$ m, near the right-rear

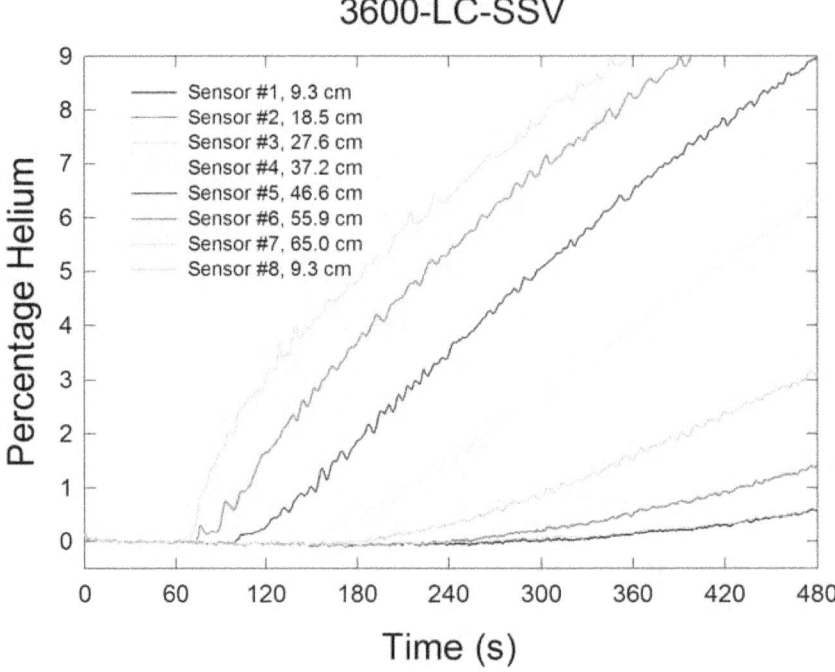

Figure 20. Helium volume percent measurements recorded at eight locations within the ¼-scale garage equipped with a single 2.4 cm × 2.4 cm vent are shown for the initial period of a one hour helium release near the floor at the center of the garage.

corner, (x,y) = (141 cm, 134.5 cm), of the enclosure. Comparison of the measured helium volume fractions for sensors #1 and #8 shows that they are nearly identical over the entire period. Similar agreement with the corresponding vertical array sensor readings was found when sensor #8 was located at different sensor heights and horizontal positions. Significant differences were observed only when sensor #8 was located immediately above the helium release point or near a vent. These observations indicate that over most of the enclosure volume there was very little lateral variation in helium volume fraction and that the measurements along the vertical sensor array were representative of the vertical concentration distributions elsewhere in the enclosure.

Inspection of Figure 19 shows that the helium concentrations began to rise shortly after the flow was initiated at 60 s and that the helium levels continued to increase at each height until the helium flow was halted at 3660 s. During the release period a vertical stratification of the helium concentration developed, with higher helium levels near the ceiling. At the end of the release period helium volume fractions varied systematically with height from approximately 35 % at z = 0.093 m to 46 % at z = 0.650 m. When the helium flow was halted, the helium volume fractions at the highest measurement locations began to drop almost immediately, while those at the lower positions continued to rise for varying periods before beginning to fall. A result of this behavior was that during the post-release period the helium volume fractions approached each other quickly and eventually began to decrease at roughly constant rates. Even so, a weak stratification remained, with the helium volume fraction at 32,400 s varying from roughly 27 % to 27.9 % for the lowest to highest sensor locations.

Additional insights concerning the concentration time behaviors are obtained by expanding the time and helium volume fraction ranges of the data shown in Figure 19. Figure 20 shows the initial eight minutes of the concentration time records on an expanded helium concentration range. Recalling that the helium flow did not start until 60 s, it is evident that helium reached sensor #7, located 0.65 m above the floor, very shortly after the flow started. The time when helium was first detected is characterized for a given sensor by defining t_{init} as the difference between the time helium was first detected and the time when the helium flow was started. It is clear from the helium concentration profiles that helium was

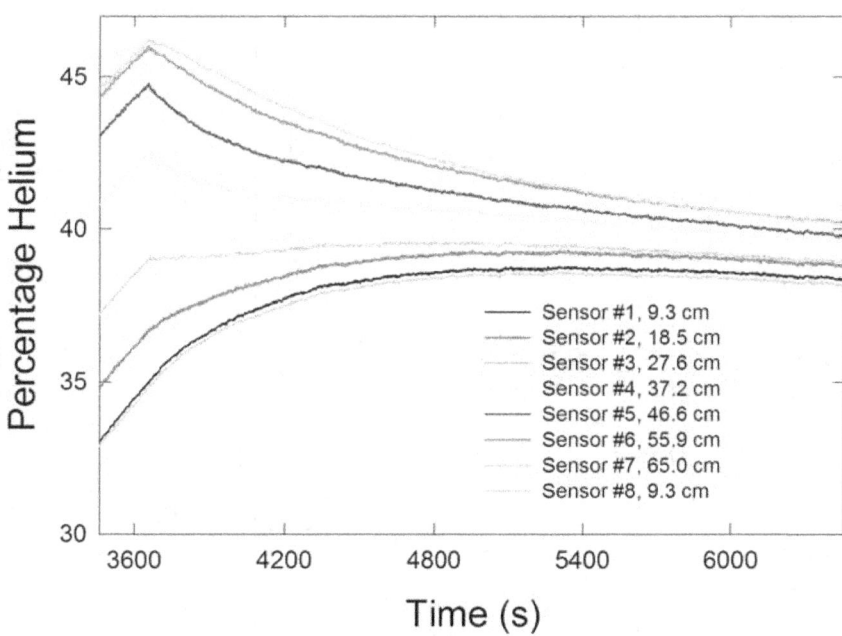

3600-LC-SSV

Sensor #1, 9.3 cm
Sensor #2, 18.5 cm
Sensor #3, 27.6 cm
Sensor #4, 37.2 cm
Sensor #5, 46.6 cm
Sensor #6, 55.9 cm
Sensor #7, 65.0 cm
Sensor #8, 9.3 cm

Figure 21. Helium volume percent measurements recorded at eight locations within the ¼-scale garage equipped with a single 2.4 cm × 2.4 cm vent are shown for a period ranging from 200 s before to 2800 s after the flow was halted for a one hour helium release near the floor at the center of the garage.

filling the enclosure from the top down since t_{init} increased with decreasing sensor height. Values of t_{init} were estimated from the data for each of the eight sensors and are included in Table 3, which summarizes a number of quantitative measures used to characterize the concentration time behaviors. For the shortest times the uncertainties in t_{init} are on the order of 1 s. The uncertainties increased somewhat for the lower sensors since the initial helium concentration increases were not as rapid. An uncertainty of 5 s was estimated for the t_{init} of sensor #1.

A blown-up section of Figure 19 extending from 200 s before the end of the helium release to 2800 s afterwards is shown in Figure 21. The different time behaviors of the concentration at the upper and lower locations discussed earlier are evident. As a means for quantitatively characterizing these different time behaviors, values of the helium volume fraction at 3600 s following the beginning of the release, $V\%_{t=3600}$, are compared with the maximum helium volume fractions measured at a given location during the entire experiment, $V\%_{max}$. The times when the values of $V\%_{max}$ occurred, $t_{V\%max}$ were also estimated visually using the curves shown in Figure 21. Values of $V\%_{t=3600}$, $V\%_{max}$, and $t_{V\%max}$ for each of the eight sensor locations are included in Table 3 for test 3600-LC-SSV.

It is evident in Figure 21 and from the quantitative values in Table 3 that the vertical concentration gradient varied with height. Near the ceiling the concentration was nearly uniform. As the measurement position moved downward the gradient became larger. The largest change in concentration occurred between sensors #3 and #4. Substantial, but somewhat smaller, concentration gradients were present between the sensors located lower in the enclosure.

The plots show that the helium concentrations for the four highest sensor locations began to decrease as soon as the helium flow was shut off, while the volume fractions at the lower locations continued to increase for substantial periods of time. The peak concentrations at the upper sensors were sharp, and the uncertainty in time is probably only a second or two. As the measurement position moved closer to the floor, the time required to reach a maximum concentration and the relative percentage increase both increased on going down from sensor #3 to sensor #1. The helium volume fraction at the

Table 3. Summary of Quantitative Measures for Eighteen Test Conditions.

3600-LC-SSV

	Sensor #1	Sensor #2	Sensor #3	Sensor #4	Sensor #5	Sensor #6	Sensor #7	Sensor #8
t_{init} (s)	196	130	104	60	39	14	8	196
$V\%_{t=3600}$	35.0	36.7	39.0	42.6	44.8	46.0	46.2	34.8
$V\%_{max}$	38.7	39.2	39.5	42.6	44.8	46.0	46.2	38.5
$t_{V\%max}$ (s)	5300	5100	4800	3598	3599	3598	3598	5300
$V\%_{t=7200}$	37.9	38.3	38.4	39.0	39.2	39.6	39.5	37.7
$V\%$ Slope$_{t=7200}$ (s^{-1})	-0.00053	-0.00054	-0.00054	-0.00059	-0.00061	-0.00063	-0.00066	-0.00054

3600-LC-SLV

	Sensor #1	Sensor #2	Sensor #3	Sensor #4	Sensor #5	Sensor #6	Sensor #7	Sensor #8
t_{init} (s)	230	182	120	82	43	26	14	230
$V\%_{t=3600}$	35.3	37.0	39.3	42.8	45.0	46.2	46.4	35.1
$V\%_{max}$	37.5	38.3	39.3	42.8	45.0	46.2	46.4	37.3
$t_{V\%max}$ (s)	4500	4500	3610	3600	3600	3600	3600	4500
$V\%_{t=7200}$	35.3	35.9	36.3	37.2	37.6	38.1	38.2	35.0
$V\%$ Slope$_{t=7200}$ (s^{-1})	-0.00117	-0.00113	-0.00122	-0.00124	-0.00131	-0.00132	-0.00138	-0.00111

3600-LC-ULV

	Sensor #1	Sensor #2	Sensor #3	Sensor #4	Sensor #5	Sensor #6	Sensor #7	Sensor #8
t_{init} (s)	235	164	115	68	44	19	8	2
$V\%_{t=3600}$	18.0	21.9	25.8	30.9	34.1	35.5	35.7	36.9
$V\%_{max}$	18.1	21.9	25.8	30.9	34.1	35.5	35.7	36.9
$t_{V\%max}$ (s)	3620	3610	3605	3603	3602	3600	3600	3600
$V\%_{t=7200}$	8.2	9.3	10.3	11.4	12.2	13.0	13.3	13.0
$V\%$ Slope$_{t=7200}$ (s^{-1})	-0.00133	-0.00168	-0.00197	-0.00229	-0.00250	-0.00267	-0.00291	-0.00283

3600-LR-SSV

	Sensor #1	Sensor #2	Sensor #3	Sensor #4	Sensor #5	Sensor #6	Sensor #7	Sensor #8
t_{init} (s)	188	174	114	68	26	25	14	188
$V\%_{t=3600}$	35.0	36.7	39.0	42.6	44.7	46.1	46.2	34.8
$V\%_{max}$	38.8	39.3	39.6	42.6	44.7	46.1	46.3	38.6
$t_{V\%max}$ (s)	5400	5000	4900	3600	3600	3600	3620	5400
$V\%_{t=7200}$	38.1	38.5	38.6	39.2	39.3	39.7	39.6	37.9
$V\%$ Slope$_{t=7200}$ (s^{-1})	-0.00055	-0.00058	-0.00060	-0.00063	-0.00063	-0.00068	-0.00071	-0.00056

26

3600-LR-SLV

	Sensor #1	Sensor #2	Sensor #3	Sensor #4	Sensor #5	Sensor #6	Sensor #7	Sensor #8
t_{init} (s)	205	185	140	73	27	24	15	205
$V\%_{t=3600}$	35.0	36.8	39.1	42.7	44.8	46.3	46.5	34.9
$V\%_{max}$	37.5	38.3	39.2	42.7	44.8	46.3	46.4	37.3
$t_{V\%max}$ (s)	4650	4450	3620	3600	3600	3600	3620	4650
$V\%_{t=7200}$	-	-	-	-	-	-	-	-
$V\%$ Slope$_{t=7200}$ (s^{-1})	-	-	-	-	-	-	-	-

3600-LR-ULV

	Sensor #1	Sensor #2	Sensor #3	Sensor #4	Sensor #5	Sensor #6	Sensor #7	Sensor #8
t_{init} (s)	194	152	83	69	28	25	16	2
$V\%_{t=3600}$	17.5	21.2	25.7	30.9	34.3	35.9	36.0	37.0
$V\%_{max}$	17.5	21.2	25.6	30.9	34.3	35.9	36.0	37.0
$t_{V\%max}$ (s)	3610	3615	3602	3604	3604	3604	3620	3600
$V\%_{t=7200}$	8.3	9.3	10.3	11.4	12.2	13.0	13.3	13.0
$V\%$ Slope$_{t=7200}$ (s^{-1})	-0.00134	-0.00160	-0.00190	-0.00224	-0.00249	-0.00265	-0.00284	-0.00274

3600-UC-SSV

	Sensor #1	Sensor #2	Sensor #3	Sensor #4	Sensor #5	Sensor #6	Sensor #7	Sensor #8
t_{init} (s)	400	300	216	128	89	45	21	400
$V\%_{t=3600}$	31.8	33.7	36.5	41.2	46.2	51.8	56.9	31.6
$V\%_{max}$	40.1	40.7	41.0	42.7	46.5	51.8	56.9	39.9
$t_{V\%max}$ (s)	5850	5800	5590	4260	3760	3610	3610	5850
$V\%_{t=7200}$	39.6	40.1	40.2	40.9	41.1	41.5	41.4	39.4
$V\%$ Slope$_{t=7200}$ (s^{-1})	-0.00051	-0.00053	-0.00059	-0.00064	-0.00067	-0.00076	-0.00078	-0.00051

3600-UC-SLV

	Sensor #1	Sensor #2	Sensor #3	Sensor #4	Sensor #5	Sensor #6	Sensor #7	Sensor #8
t_{init} (s)	370	254	205	134	87	47	20	370
$V\%_{t=3600}$	31.5	33.4	36.1	41.0	46.0	51.6	56.7	31.3
$V\%_{max}$	38.1	38.8	39.6	42.0	46.2	51.6	56.7	37.9
$t_{V\%max}$ (s)	5320	5150	4990	4110	3750	3625	3608	5320
$V\%_{t=7200}$	36.7	37.4	37.7	38.7	39.1	39.6	39.6	36.5
$V\%$ Slope$_{t=7200}$ (s^{-1})	-0.00099	-0.00104	-0.00109	-0.00119	-0.00123	-0.00138	-0.00137	-0.00105

3600-UC-ULV

	Sensor #1	Sensor #2	Sensor #3	Sensor #4	Sensor #5	Sensor #6	Sensor #7	Sensor #8
t_{init} (s)	425	300	205	126	74	42	23	9
$V\%_{t=3660}$	10.9	13.0	15.6	19.5	24.0	29.8	36.5	41.1
$V\%o_{max}$	10.9	13.0	15.6	19.5	24.0	29.8	36.5	41.1
$t_{V\%o\,max}$ (s)	3625	3610	3630	3610	3620	3605	3607	3604
$V\%_{t=7200}$	6.9	7.7	8.5	9.4	10.0	10.6	10.8	11.0
$V\%$ Slope$_{t=7200}$	-0.00107	-0.00129	-0.00153	-0.00177	-0.00196	-0.00215	-0.00222	-0.00226

14400-LC-SSV

	Sensor #1	Sensor #2	Sensor #3	Sensor #4	Sensor #5	Sensor #6	Sensor #7	Sensor #8
t_{init} (s)	350	260	200	120	81	36	18	320
$V\%_{t=14400}$	37.4	37.8	38.2	39.5	40.2	41.2	41.5	37.2
$V\%o_{max}$	37.6	38.1	38.4	39.5	40.2	41.2	41.5	37.4
$t_{V\%o\,max}$ (s)	15130	14900	14660	14400	14400	14400	14400	15130
$V\%_{t=18000}$	36.3	36.7	36.8	37.4	37.5	37.9	37.8	36.2
$V\%$ Slope$_{t=18000}$ (s^{-1})	-0.00059	-0.00058	-0.00061	-0.00059	-0.00062	-0.00065	-0.00062	-0.00058

14400-LC-SLV

	Sensor #1	Sensor #2	Sensor #3	Sensor #4	Sensor #5	Sensor #6	Sensor #7	Sensor #8
t_{init} (s)	295	260	192	115	76	36	16	290
$V\%_{t=14400}$	34.5	35.1	35.7	37.1	38.1	39.2	39.6	34.2
$V\%o_{max}$	34.5	35.2	35.7	37.1	38.1	39.2	39.6	34.2
$t_{V\%o\,max}$ (s)	14940	14710	14570	14410	14407	14400	14405	14940
$V\%_{t=18000}$	32.0	32.6	32.8	33.6	33.9	34.3	34.4	31.9
$V\%$ Slope$_{t=18000}$ (s^{-1})	-0.00082	-0.00085	-0.00091	-0.00092	-0.00091	-0.00102	-0.00097	-0.00082

14400-LC-ULV

	Sensor #1	Sensor #2	Sensor #3	Sensor #4	Sensor #5	Sensor #6	Sensor #7	Sensor #8
t_{init} (s)	326	262	182	114	82	36	28	7
$V\%_{t=14400}$	9.2	10.8	12.4	14.4	15.8	16.7	17.0	17.6
$V\%o_{max}$	9.2	10.8	12.4	14.4	15.8	16.7	17.0	17.6
$t_{V\%o\,max}$ (s)	14420	14430	14415	14405	14400	14405	14410	14400
$V\%_{t=18000}$	5.4	6.1	6.6	7.2	7.6	8.0	8.1	8.0
$V\%$ Slope$_{t=18000}$ (s^{-1})	-0.00077	-0.00091	-0.00101	-0.00118	-0.00124	-0.00138	-0.00138	-0.00135

14400-LR-SSV

	Sensor #1	Sensor #2	Sensor #3	Sensor #4	Sensor #5	Sensor #6	Sensor #7	Sensor #8
t_{init} (s)	330	304	179	120	98	46	29	330
$V\%_{t=14400}$	36.6	37.1	37.4	38.5	39.3	40.4	40.9	36.4
$V\%_{max}$	36.9	37.4	37.6	38.7	39.4	40.4	40.9	36.7
$t_{V\%max}$ (s)	15230	15085	14965	14580	14438	14406	144014	15230
$V\%_{t=18000}$	35.7	36.1	36.1	36.7	36.8	37.2	37.1	35.5
$V\%$ Slope$_{t=18000}$ (s^{-1})	-0.00061	-0.00057	-0.00063	-0.00062	-0.00061	-0.00066	-0.00062	-0.00056

14400-LR-SLV

	Sensor #1	Sensor #2	Sensor #3	Sensor #4	Sensor #5	Sensor #6	Sensor #7	Sensor #8
t_{init} (s)	385	260	210	122	82	48	31	260
$V\%_{t=14400}$	34.4	35.1	35.6	37.1	38.0	39.0	39.3	34.2
$V\%_{max}$	34.4	35.1	35.6	37.1	38.0	39.0	39.3	34.3
$t_{V\%max}$ (s)	14695	14645	14540	14400	14403	14405	14414	14695
$V\%_{t=18000}$	31.9	32.5	32.7	33.6	33.8	34.2	34.3	31.8
$V\%$ Slope$_{t=18000}$ (s^{-1})	-0.00089	-0.00089	-0.00096	-0.00100	-0.00100	-0.00101	-0.00102	-0.00087

14400-LR-ULV

	Sensor #1	Sensor #2	Sensor #3	Sensor #4	Sensor #5	Sensor #6	Sensor #7	Sensor #8
t_{init} (s)	328	298	229	124	50	46	31	5
$V\%_{t=14400}$	8.9	10.4	12.1	14.2	15.5	16.5	16.7	17.4
$V\%_{max}$	8.9	10.4	12.1	14.2	15.5	16.5	16.7	17.4
$t_{V\%max}$ (s)	14418	14416	14400	14402	14408	14407	14401	14400
$V\%_{t=18000}$	5.4	5.9	6.5	7.1	7.5	7.9	8.0	8.0
$V\%$ Slope$_{t=18000}$ (s^{-1})	-0.00070	-0.00084	-0.00095	-0.00109	-0.00119	-0.00128	-0.00135	-0.00131

14400-UC-SSV

	Sensor #1	Sensor #2	Sensor #3	Sensor #4	Sensor #5	Sensor #6	Sensor #7	Sensor #8
t_{init} (s)	525	460	328	225	145	83	41	460
$V\%_{t=14400}$	36.6	37.0	37.4	38.7	39.7	41.3	42.7	36.4
$V\%_{max}$	37.2	37.7	38.0	39.1	39.9	41.4	42.8	37.0
$t_{V\%max}$ (s)	15640	15400	15300	14920	14620	14470	14422	15690
$V\%_{t=18000}$	36.3	36.7	36.7	37.3	37.4	37.8	37.7	36.1
$V\%$ Slope$_{t=18000}$ (s^{-1})	-0.00048	-0.00051	-0.00050	-0.00052	-0.00055	-0.00056	-0.00060	-0.00049

14400-UC-SLV

	Sensor #1	Sensor #2	Sensor #3	Sensor #4	Sensor #5	Sensor #6	Sensor #7	Sensor #8
t_{init} (s)	490	390	292	170	118	70	44	490
$V\%_{t=14400}$	34.0	34.7	35.3	36.9	38.2	39.9	41.4	33.9
$V\%_{max}$	34.2	35.0	35.6	37.1	38.4	40.0	41.4	34.1
$t_{V\%max}$ (s)	15190	15090	14930	14695	14540	14452	14420	15190
$V\%_{t=18000}$	32.1	32.7	33.0	33.9	34.2	34.6	34.7	32.1
$V\%$ Slope$_{t=18000}$ (s^{-1})	-0.00091	-0.00088	-0.00099	-0.00098	-0.00102	-0.00105	-0.00110	-0.00087

14400-UC-ULV

	Sensor #1	Sensor #2	Sensor #3	Sensor #4	Sensor #5	Sensor #6	Sensor #7	Sensor #8
t_{init} (s)	540	465	305	204	124	82	40	21
$V\%_{t=14400}$	7.2	8.2	9.4	11.0	12.7	14.8	17.0	19.6
$V\%_{max}$	7.2	8.2	9.4	11.0	12.7	14.8	17.0	19.6
$t_{V\%max}$ (s)	14410	14420	14430	14440	14439	14438	14423	14405
$V\%_{t=18000}$	5.0	5.5	5.9	6.5	6.8	7.1	7.3	7.3
$V\%$ Slope$_{t=18000}$ (s^{-1})	-0.00055	-0.00065	-0.00077	-0.00085	-0.00096	-0.00103	-0.00110	-0.00107

sensor #1 location continued to increase for roughly 1650 s, increasing by nearly 11 % following the end of the release. The uncertainties in $t_{V\%max}$ at the lower sensor locations were considerably higher due to the relatively slow changes in concentration. For sensors #1 and #8 the uncertainty was estimated as ± 200 s.

Following the end of a release, helium concentrations began to drop as helium diffused or flowed through the vent and was replaced by air. Values of the helium concentration at 7200 s following the release start, $V\%_{t=7200}$ and the slope for the concentration fall off, $V\% Slope_{t=7200}$, are used as quantitative measures to characterize the concentration decay for a one hour release. Values for these two parameters are included in Table 3 for the one hour, lower-center release with the small centered vent. Uncertainties are equal to or less than the least significant figures reported. The results reveal that a small vertical concentration gradient remained one hour into the post-release period. The slopes of the concentration profiles were similar, but concentrations at the higher sensor positions were falling off slightly faster. This indicates that the average vertical concentration gradient was continuing to decrease at this time.

The plot of differential pressure included in Figure 19 shows that the pressure increased from zero to roughly 0.22 Pa when the flow was initiated. Since the flow from the opening was primarily air at this time, and the volume flow rate through the opening was equal to the helium flow rate, it should be possible to predict the pressure increase using the results of the fan test described earlier for the enclosure with a single small vent. Eq. (2) and the parameters included in Table 2 were used. The result of the calculation was ΔP = 0.342 Pa. This value is over 50 % higher than observed, but it should be kept in mind that the fan measurements were recorded for much larger pressure drops, and the curve has been extrapolated down into this volume flow rate range.

During the helium release the pressure drop across the opening decreased by 0.04 Pa over the hour. This pressure change is likely due to two counteracting effects. The first is the reduction in differential pressure associated with decreasing gas density in the garage as the helium concentration builds up. Sensor #4 is located at the same height as the vent in the front face. Assuming the helium concentration at the vent was the same as at sensor #4, the helium volume fraction in the vent at the end of the release was 42.6 % (see Table 3). This means the density of the gas had decreased by 37 %. According to Eq. (1), the ΔP due to the flow should vary with the square root of density. As a result, a 20 % drop in ΔP is expected as compared to when air was flowing through the opening at the start of the release. When the flow was halted, the differential pressure dropped from 0.18 Pa to 0.04 Pa, indicating that the ΔP due to the flow was 0.14 Pa, which is a 64 % reduction. The difference suggests that an additional effect was affecting ΔP.

The second effect is more subtle. It becomes evident as the small positive differential pressure present following the end of the helium release. As the helium flowed into the garage the density inside the enclosure decreased, creating a vertical hydrostatic pressure difference across the opening between the interior and outside. The pressure difference was positive at the top of the vent and negative at the bottom. The total differential pressure change across the opening can be computed as the difference in density times the acceleration of gravity times the change in height across the opening. The result using the helium concentration for sensor #4 to determine the interior density is 0.106 Pa. The small positive differential pressure evident during the early post-release period indicates that the pressure sensor was located above the neutral plane for this pressure gradient. As helium was lost from the garage and replaced with air during the post-release period, the gas density increased and the remaining hydrostatic pressure difference decreased. Eventually, the differential pressure across the vent fell to zero.

A 29.6 L/min (0.000493 m³/s) flow of air was used to sweep remaining helium from the enclosure at the end of an experiment. Figure 22 shows a plot of the differential pressure for 3600-LC-SSV over a period from just prior to the start of the air flow until the flow was halted and the baseline recorded. By averaging over brief periods just prior to and after the air flow was stopped, the ΔP due to the air flow was determined to be -0.983 Pa. The predicted value of ΔP using the measured flow rate, Eq. (2), and the coefficients in Table 2 is 1.21 Pa. This value is 23 % higher than observed. Earlier, it was found that for a volume flow rate of 0.000249 m³/s the same approach overestimated ΔP by more than 50 %. This suggests that the predictions become less accurate as ΔP become smaller.

Values of measured differential pressure changes for 3600-LC-SSV recorded when the helium flow was initiated and halted and when the air flow was halted are summarized in Table 4.

31

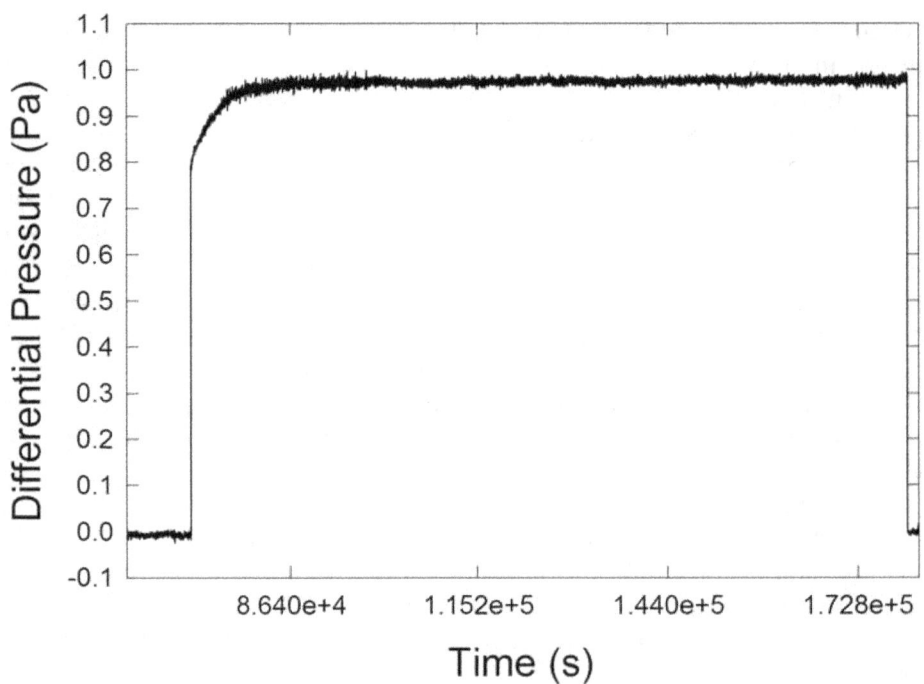

Figure 22. The differential pressure is shown for 3600-LC-SSV over the time from just before to just after a 29.6 L/min air flow was used to sweep remaining helium from the garage.

Table 4. Variations in Differential Pressure Associated with Changes in Flow Conditions (ΔP)

Test	Start Helium Flow	Halt Helium Flow	Halt Air Flow
3600-LC-SSV	0.22 Pa	-0.14 Pa	-0.983 Pa
3600-LC-SLV	--	--	--
3600-LC-ULV	0.08 Pa	-0.55 Pa	-0.315 Pa
3600-LR-SSV	0.23 Pa	-0.13 Pa	-0.982 Pa
3600-LR-SLV	0.09 Pa	-0.08 Pa	-0.360 Pa
3600-LR-ULV	0.08 Pa	-0.55 Pa	-0.317 Pa
3600-UC-SSV	0.23 Pa	-0.14 Pa	-0.989 Pa
3600-UC-SLV	0.09 Pa	-0.07 Pa	-0.368 Pa
3600-UC-ULV	0.08 Pa	-0.50 Pa	-0.317 Pa
14400-LC-SSV	0.025 Pa	-0.020 Pa	-0.984 Pa
14400-LC-SLV	--	-0.025 Pa	-0.363 Pa
14400-LC-ULV	0.005 Pa	-0.085 Pa	-0.323 Pa
14400-LR-SSV	0.025 Pa	-0.02 Pa	-0.973 Pa
14400-LR-SLV	0.005 Pa	-0.02 Pa	-0.363 Pa
14400-LR-ULV	0.01 Pa	-0.09 Pa	-0.316 Pa
14400-UC-SSV	--	-0.02 Pa	-0.987 Pa
14400-UC-SLV	0.01 Pa	-0.02 Pa	-0.373 Pa
14400-UC-ULV	0.01 Pa	-0.09 Pa	-0.323 Pa

Figure 23 shows the concentration profiles for a one hour helium release from the lower center position into the garage equipped with a single 3.05 cm × 3.05 cm vent in the front wall. Sensor #8 was located in the same position as for 3600-LC-SSV. Due to a technical problem, the corresponding differential pressure was not recorded and is not shown. Comparison with the corresponding profiles for the enclosure with the single small vent in Figure 19 shows the helium volume percent curves have similar

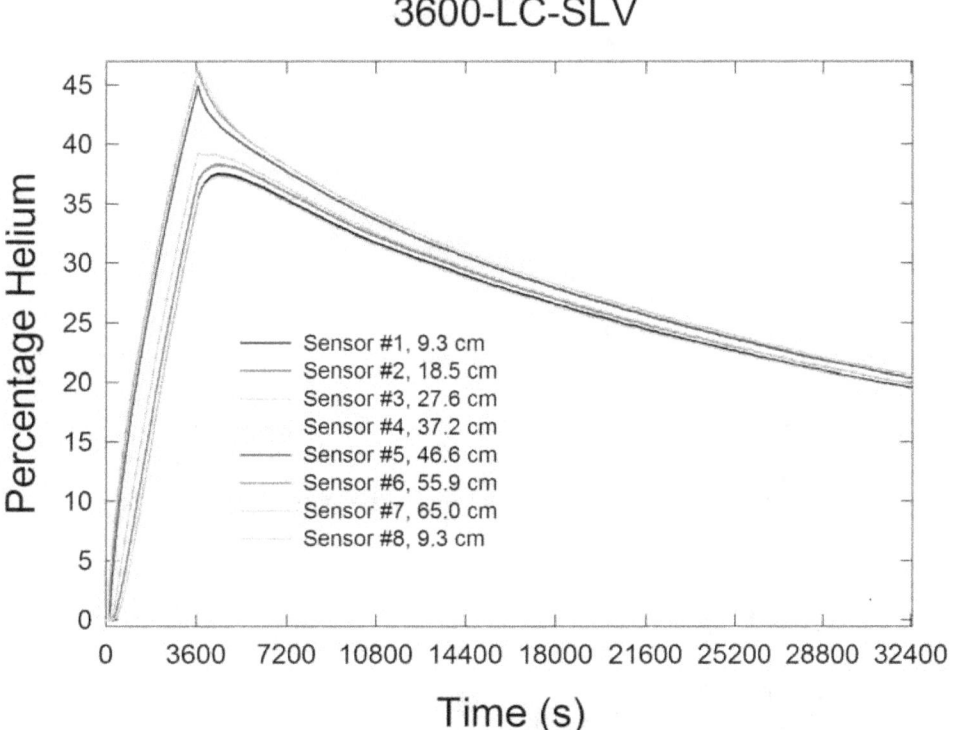

3600-LC-SLV

Figure 23. Helium volume percent measurements recorded at eight locations within the ¼-scale garage equipped with a single 3.05 cm × 3.05 cm vent are shown for the initial period of a one hour helium release near the floor at the center of the garage.

appearances and magnitudes for the two vent sizes during the release phases. However, there are apparent differences during the post-release phase. Helium concentrations fall off faster with the larger opening as indicated by the lower helium volume percentages present eight hours after the release ended.

The general behaviors of the concentration profiles for the two vent sizes are reflected in the numerical measures included in Table 3, which show that the helium volume fractions for a given height were within 1 % of each other at the ends of the release periods. The values for the experiment with the larger vent are consistently higher (0.2 % or 0.3 %), but the differences are on the order of size of those expected due to experimentally observed variations in measured helium flow rates.

Even though the helium concentration profiles were very similar for the two vent sizes during the helium release periods, the values listed in Table 3 indicate that the times when helium was first detected at the various sensors were slightly longer for the larger vent. This suggests that the initial filling time for the enclosure was increased when the larger vent was used.

The observed differences in post-release loss of helium from the enclosure were also captured by the quantitative measures. While the maximum observed helium volume percent values for the higher positions, sensors #4 to #7, were similar for both vent sizes since they occurred very close to the time the helium releases ended, those for the lower sensors were noticeably reduced with the larger vent, suggesting a more rapid loss of helium from the garage. A more rapid loss of helium is also consistent with the shorter times at which the maxima were observed at these heights for 3600-LC-SLV. This is particularly the case for sensor #3, for which $t_{v\%max}$ dropped from 4800 s to 3610 s.

The vent size effect was more pronounced in the measures used to characterize the concentration fall off at 7200 s. The observed helium volume fractions at this time for each of the sensor heights were markedly lower with the larger vent. Most telling are the observed slopes at this time. The helium concentrations were falling at rates more than twice as high for the garage with the larger vent despite the fact that the absolute concentrations were smaller at this time.

33

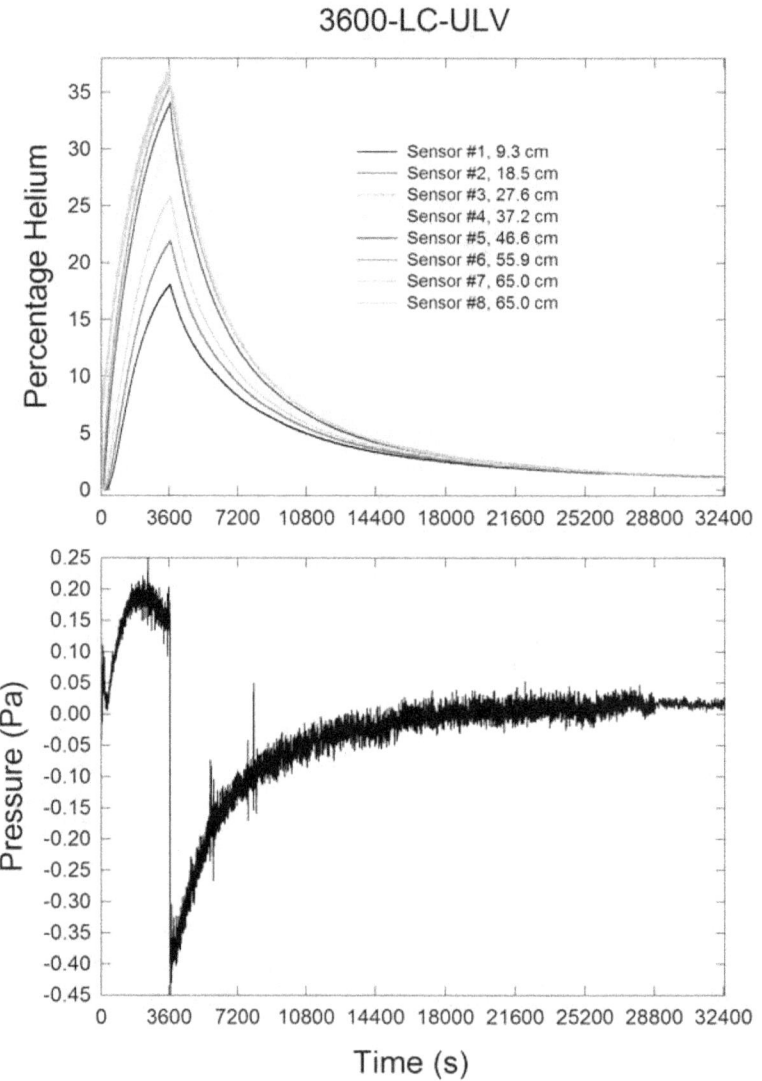

Figure 24. The upper plot shows helium volume percent as a function of time at eight locations for a one hour release of helium from the lower center position into the ¼-scale garage with 2.15 cm × 2.15 cm upper and lower openings in the front face. The lower plot shows the differential pressure across the face.

It appears that the primary effect of increasing the vent area by 61 % was to greatly increase the loss rate of helium from the enclosure during the post-release period. The change in vent size had very little effect on concentration levels during the helium release, suggesting that very little air entered the garage during this time. Helium lost from the enclosure during this time must be due to helium/air mixture flow from the enclosure and not to exchange with the outside air.

Figure 24 shows helium concentration temporal profiles for the eight sensor locations (upper plot) and the differential pressure (lower plot) for a one hour release of helium into the garage with a front wall having two 2.15 cm × 2.15 cm vents located 2.54 cm from the floor and ceiling. For these measurements sensor #8 was located directly above the helium source at the same height as sensor #7, i.e., $(x,y,z) = (0.750$ m, 0.750 m, 0.650 m). Comparison of Figure 24 with Figure 19 and Figure 23 shows that the concentration profiles for the seven sensors forming the vertical array had very different temporal profiles than observed for the comparable experiments with single vents in the center of the face. During the helium release period the maximum observed concentrations were reduced considerably, and the vertical stratification was

34

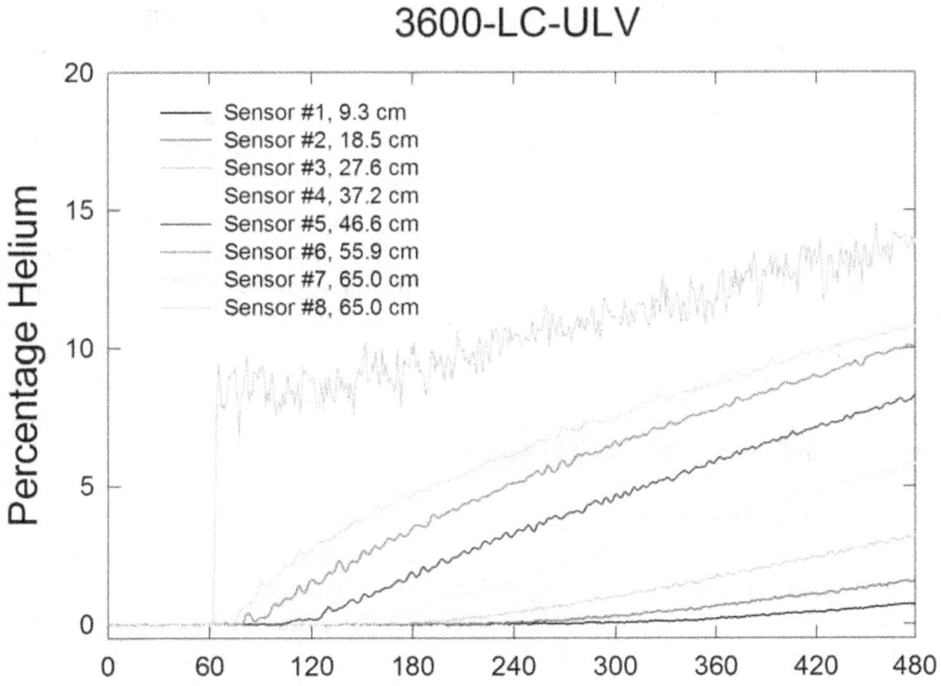

3600-LC-ULV

Figure 25. Helium volume percent measurements recorded at eight locations within the ¼-scale garage equipped with upper and lower 2.15 cm × 2.15 cm vents are shown for the initial period of a one hour helium release near the floor at the center of the garage.

increased as indicated by a larger average concentration gradient. Unlike for the cases with single central vents, significant vertical concentration gradients extended into the upper layer over sensors #5, #6, and #7. These conclusions are supported by the values of $V\%_{t=3600}$ included in Table 3 for the three flow conditions. These observations suggest that not only were higher helium concentrations flowing from the upper vent, but also that some air was entering the enclosure at the lower vent during the release period.

The helium volume fractions recorded by sensor #8 just prior to the end of the helium release were roughly 1 % higher and showed larger fluctuations than those recorded at the same height by sensor #7 at a location outside of the helium plume. However, the relatively small difference indicates that large amounts of air were mixed with the helium released from the burner over the 0.443 m (12.3 burner diameters) distance between the helium source and this measurement location. The large amount of mixing is also evident in the initial rise of helium concentration when the release was initiated at 60 s. Figure 25 shows helium concentration plots for the eight locations over the initial seven minutes of the flow. Helium reached sensor #8 within two seconds and quickly attained a plateau volume fraction value of 8.5 %. Since only air was present outside of the plume at this time, this demonstrates that the helium flow along the centerline was diluted with air by more than a factor of ten during this initial mixing phase. It is clear from Figure 24 that the centerline helium concentration at sensor #8 continuously increased during the release period as helium built up elsewhere in the enclosure.

The times when helium was first detected by the various sensors are evident in Figure 25 and are quantified in Table 3. The periods required for the two-vent face were very similar to those observed with the single 3.05 cm × 3.05 cm vent. Since the total vent areas of for these two experiments were similar, this suggests that the initial filling time of the enclosure was determined primarily by the total vent area and was independent of vent configuration.

The data shown in Figure 24 indicates that the helium concentration at sensor #8 dropped immediately to a value similar to that at sensor #7 when the helium flow was halted. As the values of $V\%_{max}$ included in Table 3 for 3600-LC-ULV indicate, the helium concentrations for each of the sensors on the vertical array began to drop within 20 s following the end of the release. This behavior is very different than observed for the cases with a single vent, for which the concentrations at the lower sensor positions continued to increase for significant periods.

Comparisons of the helium concentration profiles in Figure 24 with those in Figure 19 and Figure 23 for the post-release periods show that the helium concentrations fell off much faster when there were two vents present. It is also clear that there was more vertical stratification during the post-release phase with the two vents. Using the values of $V\%_{t=7200}$ included in Table 3 for the two-vent case, it can be shown that helium concentration decreased between 54 % and 63 % at the seven sensor locations during the one hour period following the end of the release. The relative decreases were roughly constant for the upper four sensor locations, while the relative decreases were somewhat smaller at the lower positions. In contrast, the corresponding decreases at the upper sensor locations were less than 20 %, while the concentrations at the lower locations actually increased when the larger single vent was used. These observations, along with the values for $V\%_{max}$, indicate that the helium-air exchange rate between the interior and ambient surroundings was much higher for the two-vent case despite the fact that the vent areas were comparable.

The pressure measurement shown in Figure 24 for the experiment with two vents has a complicated time dependence. When the helium flow was first started there was a barely detectable 0.08 Pa rise in the pressure. This is the pressure increase associated with air flowing from the enclosure as the helium first enters. This value of ΔP can be compared with the corresponding value of 0.22 Pa observed with a single 2.4 cm × 2.4 cm vent. The ratio of the two vent areas was 9.25 cm^2 to 5.76 cm^2. Based on Eq. (1), ΔP should vary as the area of the opening squared, so the ratio for the differential pressures changes can be estimated to be 2.6, which compares favorably with the experimental value of 2.8.

After the initial rise in differential pressure, it fell slightly for approximately 250 s, approaching zero. At this point it began to increase again, rising to a second maximum of 0.20 Pa approximately 2200 s after the start of the helium flow. At this point the differential pressure began to fall again, reaching a value of 0.15 Pa by the time the helium flow was halted at 3660 s. At this time, the differential pressure immediately dropped by 0.55 Pa to -0.40 Pa. This complicated time dependence is likely due to changes in the internal hydrostatic pressure distribution as the helium concentration inside the enclosure increases, coupled with the interior pressure increase due to the helium flow into the garage.

Since the interior of the garage communicated with the surroundings at two widely spaced vertical locations, it was the difference in interior and exterior vertical hydrostatic pressure profiles which drove flows into and out the garage. As a demonstration that hydrostatic pressure variations played an important role in the observed differential pressure behavior, the interior vertical pressure distribution was calculated immediately following the end of the helium release. Figure 26 shows plots that provide insights as to how this calculation was done. Recall that the differential pressure gauge is located at a height midway between the two vents. It is expected that a helium/air mixture will flow out of the upper vent and air will flow into the lower vent. Since the vertical helium concentration distribution was available and appeared to be independent of horizontal location, it was possible to calculate the density distribution and then use this to calculate the relative hydrostatic pressure as a function of height inside the garage. The left-top portion of Figure 26 shows the measured helium volume percent (circles) as a function of height above the floor immediately after the helium flow was halted. In order to estimate the concentration at the ceiling (square), it was assumed that the concentration remained constant with height in the upper region. An estimate for the concentration at the floor (square) was obtained from a linear extrapolation using the five lowest measured values. A simple model that assumed two piecewise linear concentration variations with height was fit to the vertical values as shown by the line included in the plot. The helium volume fraction distribution was converted to a vertical density profile using the known densities of air and helium at 21 °C. The result is shown in the upper right corner of Figure 26.

The relative hydrostatic pressure distribution inside the garage was then determined by numerically integrating the product of the density and gravitational acceleration downward from $z = 0.75$ m to the floor. The corresponding distribution exterior to the enclosure was obtained using the constant density for air. The lower left panel in Figure 26 shows the resulting relative hydrostatic pressure distributions as a

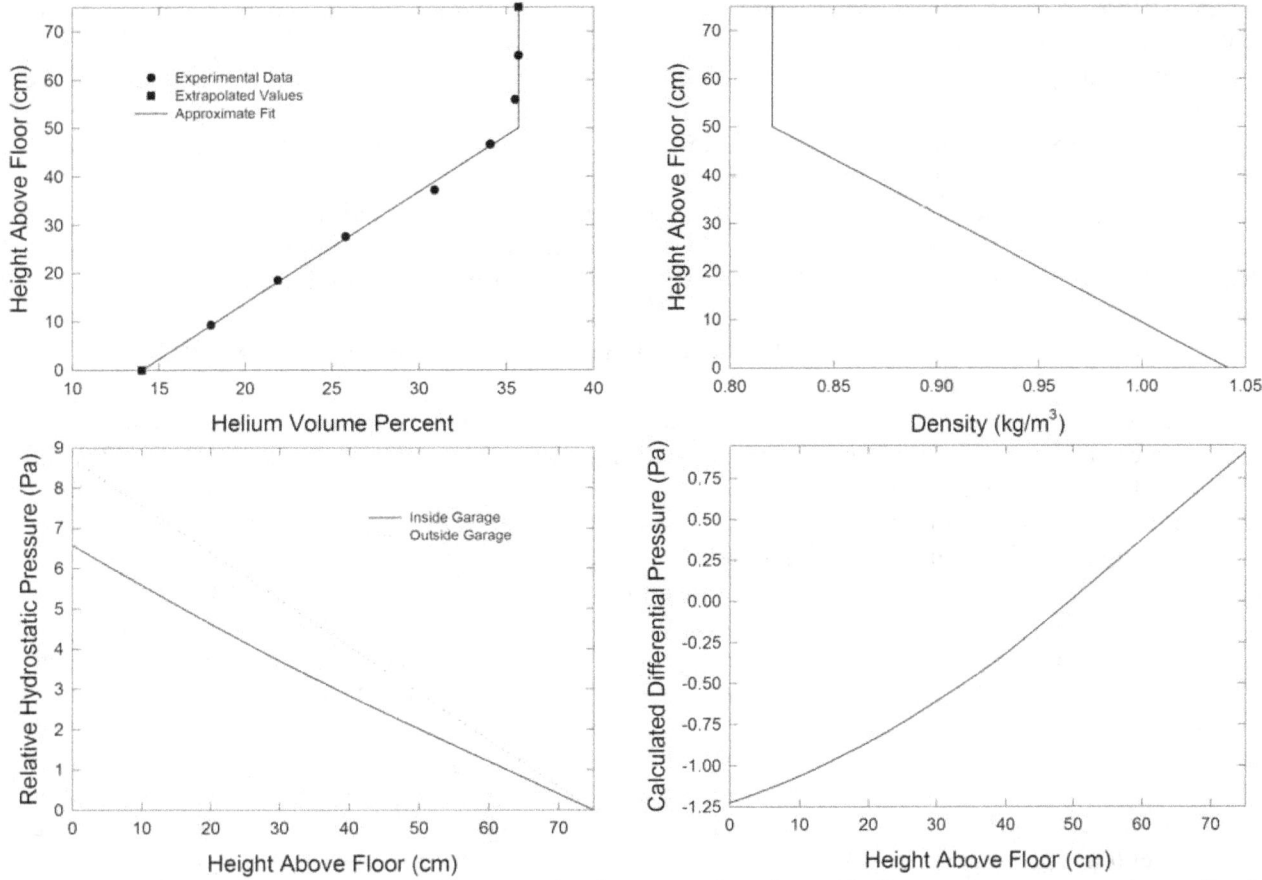

Figure 26. The four plots show height above the floor as a function of helium volume percent immediately following the end of helium release for 3600-LC-ULV (upper left), the corresponding density plot (upper right), relative vertical hydrostatic pressure difference distributions inside the garage and in the ambient (lower left), and the calculated vertical differential pressure distribution with the neutral plane located to provide the required differential pressure ratio at the vents (lower right).

function of height. The maximum difference in relative hydrostatic pressure is 2.13 Pa over the 0.75 m distance. The interior pressure distribution must adjust such that the volumetric flow rates into and out of the enclosure are equal in order to prevent the development of absolute pressure differences which would inhibit the flows. Eq. (2) indicates that in order for the flow volumes to match it is necessary for the absolute ratios of differential pressure at the vents divided by the local gas densities to be equal. The densities flowing through the upper and lower vents were assumed to be those for the helium mixture at $z = 0.725$ m (0.821 kg/m^3) and air (1.184 kg/m^3) at $z = 0.025$ cm, respectively, yielding a density ratio of 0.693. The calculated difference in hydrostatic pressures between these two heights was 2.01 Pa.

In order for the vent flows hypothesized above to have developed, it is necessary that the differential pressure be positive at the upper opening and negative at the lower. This requires that there was a location in between where the differential pressure was zero. This location is often referred to as the neutral plane height. The required location of the neutral plane was determine by calculating values of the ratio of differential pressures for the upper and lower vent locations assuming neutral plane heights covering the range of heights. The required ratio of 0.693 for the differential pressures at the upper and lower vents was found when the neutral plane was located at $z = 0.0495$ m. The lower right-hand panel in Figure 26 shows the resulting differential pressure distribution inside the enclosure when the internal hydrostatic pressure distribution was shifted to place the neutral plane at this height. The calculated differential pressure at the measurement height of $z = 0.375$ m is 0.394 Pa, which is in excellent agreement with the experimental value of 0.40 Pa.

During the post-release period the negative differential pressure rose as the helium in the garage was replaced with air, approaching zero as the helium concentrations fell to zero. This is consistent with the pressure difference at the measurement location being due to the hydrostatic pressure differences between the interior helium mixture and the exterior air.

Air was used to flush out the remaining helium at the end of the experiment. The observed value of $\Delta P = -0.315$ Pa when the air flow was halted is included in Table 4. The ratio of ΔP values for 3600-LC-SSV and 3600-LC-ULV is 2.24, which agrees well with the expected value of 2.4 determined earlier.

3.2.2. One Hour Helium Releases near the Floor at the Rear of the Garage

Figure 27 shows a plot of helium concentrations at the eight sensor locations for a one hour helium release with the burner located on the floor at the rear of garage at the center of the wall. The front wall was equipped with a single 2.4 cm × 2.4 cm opening. Sensor #8 was placed at the same height as sensor #1 near the right rear corner, $(x,y) = (141$ cm, 134.5 cm$)$. Comparison of Figure 19 and Figure 27 shows that the concentration profiles appear indistinguishable during the release period. The numerical values in Table 3 for 3600-LC-SSV and 3600-LR-SSV confirm the agreement. All of the values for $V\%_{t=3600}$ and $V\%_{max}$ fall within 0.1 % for each of the sensor locations. Values of t_{init} and $t_{V\%max}$ were also similar for the two experiments. The times for the initial detection at a given sensor vary somewhat, but the variations appear to be random, and the values for a given sensor generally fall close together.

Some systematic differences between these two experiments seemed to appear during the post-release period. The values of $V\%_{t=7200}$ for each of the sensors were systematically larger (either 0.2 % or 0.3 %) for the release in the rear of the garage. Interestingly, the slopes for the falloffs of the helium concentrations were slightly higher at this time for the rear-release case. These observations suggest that the rate of helium loss to the surroundings can vary by small amounts with time.

The pressure curves in Figure 19 and Figure 27 are also very similar. This is reflected by the close agreement in the values of differential pressure changes included in Table 4 when the helium flow was started and halted for 3600-LC-SSV and 3600-LR-SSV. The values of ΔP when the air flow was stopped also agree well.

Temporal profiles of concentration and differential pressure are shown in Figure 28 for a one hour release from the lower rear of the garage equipped with a single large vent. Sensor #8 was located at the same lower rear location as for the other single-vent cases discussed thus far. Due to a computer malfunction, only the initial 5208 seconds of the experiment were recorded. Comparisons with the corresponding results for 3600-LC-SSV, 3600-LC-SLV, and 3600-LR-SSV shown in Figure 19, Figure 23, and Figure 27 show that the concentration behaviors at the various sensor locations were very similar for the helium release phases. In fact, the quantitative results for $V\%_{t=3600}$ listed in Table 3 indicate that, with the exception of 3600-LC-SLV, the helium concentrations measured at the end of the release periods for all of the sensor locations agree within 0.2 %, with the vast majority within 0.1%. As already discussed, values for 3600-LC-SLV fell slightly higher, but were still very similar to those observed in the other three tests. The results provide strong evidence that the development of the concentration profiles during one hour helium releases with single vents in the center of the front face are independent of the vent area (5.76 cm^2 and 9.30 cm^2) and horizontal release location (20.7 cm above the floor at center and rear of the garage).

For releases at the lower center position it was found that more helium was lost from the enclosure during the post-release period with the larger single vent. The values of $V\%_{max}$ for the eight sensor locations are nearly identical for 3600-LC-SLV and 3600-LR-SLV. The times of the maximum concentrations are also very similar with those at sensors #1 and #2 occurring at earlier times as compared to 3600-LC-SSV and 3600-LR-SSV. As for 3600-LC-SLV, the helium concentration for sensor #3 reached a maximum very close to the time the helium release ended. This maximum occurred much later for tests

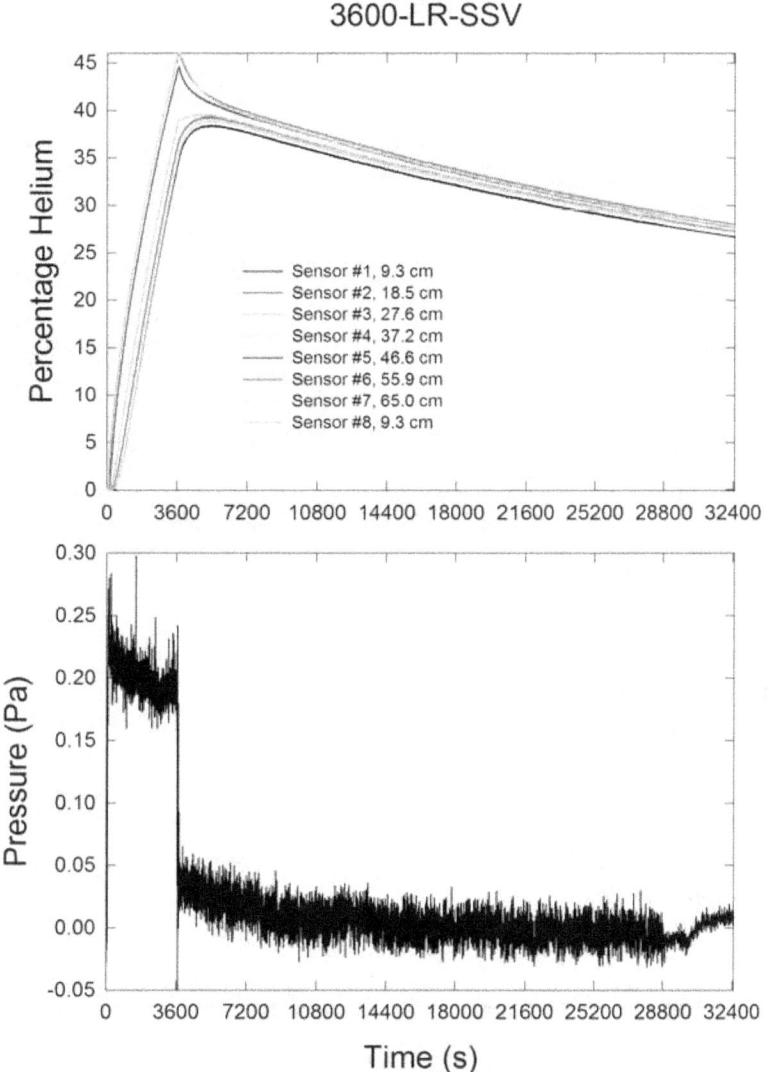

Figure 27. The upper plot shows helium volume percent as a function of time at eight locations for a one hour release of helium from the lower rear position into the ¼-scale garage with a single centered 2.4 cm × 2.4 cm opening in the front face. The lower plot shows the differential pressure across the face.

with the smaller vent. Even though $V\%_{t=18000}$ and $V\% \ Slope_{t=18000}$ values are unavailable, it is clear that the loss of helium from the garage for 3600-LR-SLV is consistent with earlier observations.

Pressure data for 3600-LC-SLV was unavailable. The differential pressure time dependence for 3600-LR-SLV is included in Figure 28, and the resulting quantitative values are included in Table 4. Table 4 also includes the differential pressure drop due to the air flow used to flush the enclosure. The changes in differential pressure when the helium flow was initiated and halted were considerably smaller than when the smaller vent was used (compare Figure 28 with Figure 19 and Figure 27 and see quantitative values in Table 4). When the small vent was used the differential pressure rapidly rose to a maximum when the helium flow was started and then dropped slowly and nearly linearly as the flow continued. When the helium flow was halted, the pressure difference dropped to a slightly positive value, which then slowly decayed with time. The differential pressure behavior in Figure 28 is different. Following the initial increase when the helium flow was started, the differential pressure remained nearly constant for around

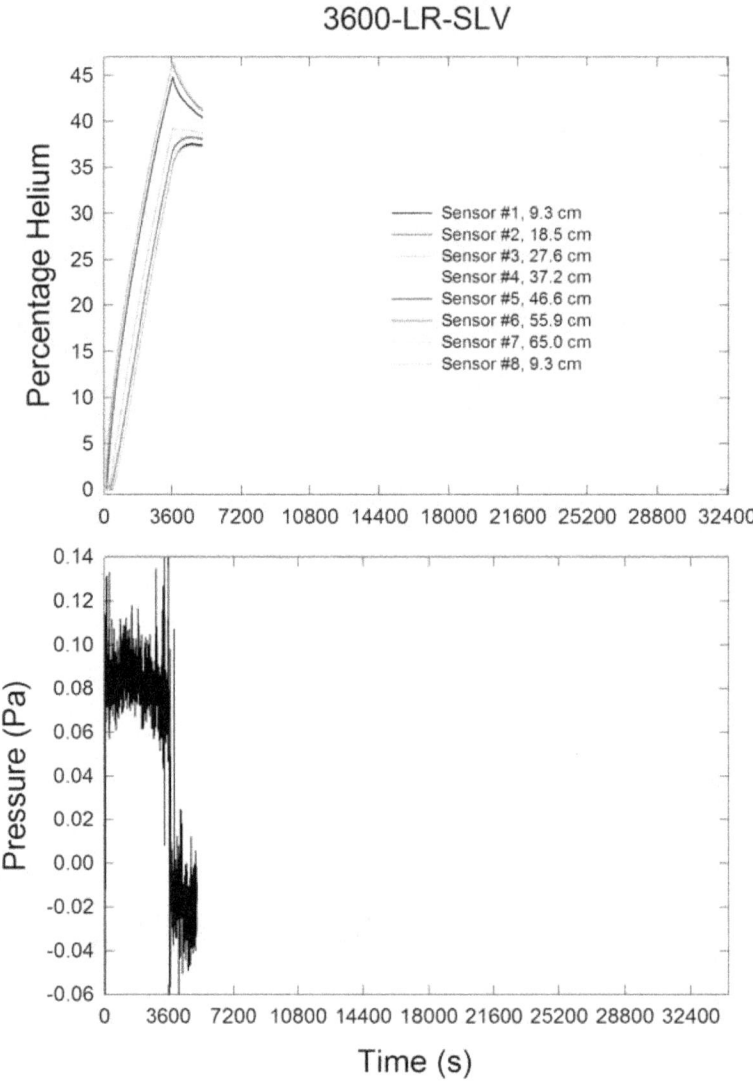

Figure 28. The upper plot shows helium volume percent as a function of time at eight locations for a one hour release of helium from the lower rear position into the ¼-scale garage with a single centered 3.05 cm × 3.05 cm opening in the front face. The lower plot shows the differential pressure across the face. Data acquisition stopped unexpectedly at 5808 s.

1800 s before dropping slightly. When the helium was shut off, the differential pressure dropped to a slightly negative value.

The total vent areas for the faces with a single large vent and the upper and lower vents are nominally identical. As a result, nearly identical differential pressure drops would be expected due to the initial helium and air flows if the flow coefficients were the same. While the values are similar for both the helium and air flows, slightly higher pressure drops were observed for the experiment with two vents. The differences amounted to 12 % to 14 %. Assuming the flow coefficients were the same for the single- and dual-vent faces, these values indicate that the total area for the two-vent system is roughly 6 % larger. This difference exceeds the uncertainties in the vent dimension measurements. Interestingly, the effective area determined by fan tests was also 6 % larger for the two-vent face (see Table 2). The higher pressure exponent for the two-vent face suggests that the difference may be due to different flow coefficients for the two faces.

40

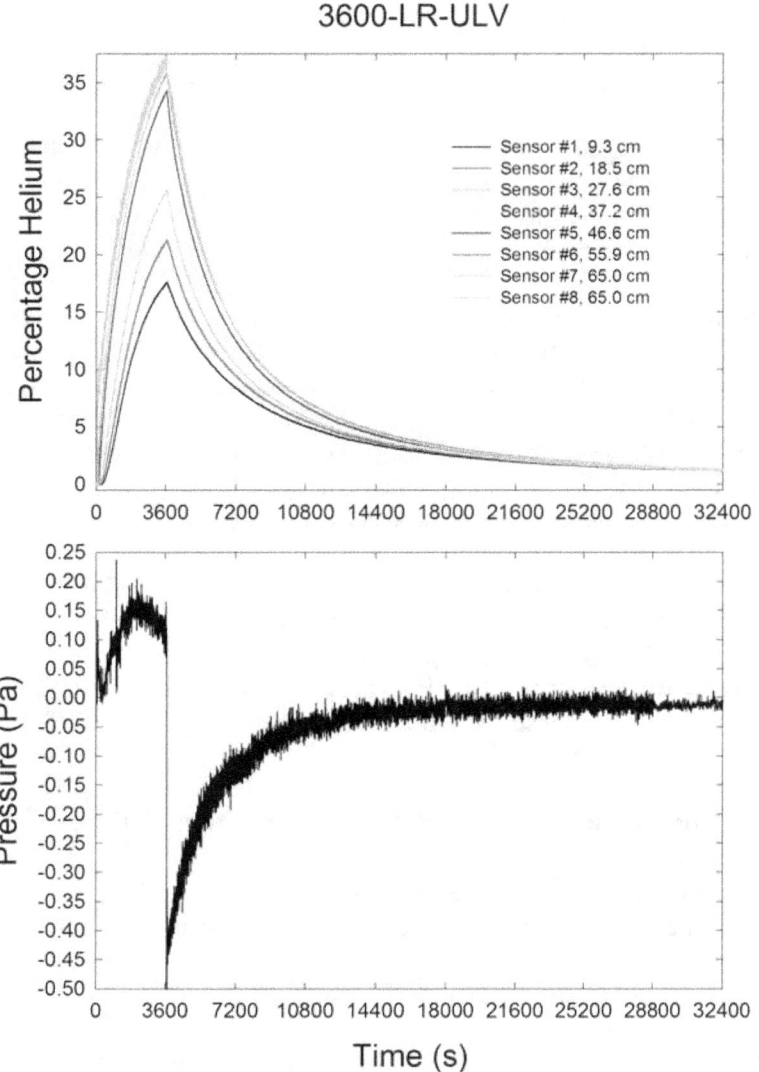

Figure 29. The upper plot shows helium volume percent as a function of time at eight locations for a one hour release of helium from the lower rear position into the ¼-scale garage with 2.15 cm × 2.15 cm upper and lower openings in the front face. The lower plot shows the differential pressure across the face.

Figure 29 shows helium concentration and differential pressure temporal profiles for a one hour helium release at the lower rear location into the garage equipped with the two-vent face. Sensor #8 was located above the release point at the same height as sensor #7. The times of initial observation of helium at the sensor locations included in Table 3 show some variations, but are consistent with the corresponding times for 3600-LC-ULV. The helium concentration profiles during the helium release are similar to those in Figure 24 for a release at the lower center of the garage with the same face. Comparison of the helium concentrations measured at the end of the release period included in Table 3 reveals a slight shift in the vertical concentration gradients, with the experiment with the lower-center release point having slightly higher concentrations in the bottom of the garage and slightly lower concentrations near the ceiling. Such a shift in vertical concentration profiles has not been observed up to now and may be an indication of a slight effect of release location on the mixing behavior for this vent configuration.

The differential pressure profile in Figure 29 is similar to that shown in Figure 24, however there are quantitative differences. The initial pressure increase due to turning on the helium flow is 0.08 Pa, which is the same as measured with burner at the center. The differential pressure then fell to near zero with a minimum around 275 s following the start of the release before rising to a second maximum of 0.16 Pa at approximately 2150 s. This maximum value is slightly less than the 0.20 Pa value observed for the center release. The pressure then fell back to 0.11 Pa by the end of the end of the helium release, which is lower by the same amount as compared to the maximum value as observed for the center release. When the helium was turned off the differential pressure dropped to -0.44 Pa for a total change of 0.55 Pa. The total change is identical to that observed for the center release, but again the pressure for the rear release is 0.04 Pa less than observed with the center release. As seen in Table 4, the pressure drops due to shutting off the air flow were nearly identical for 3600-LR-ULV and 3600-LC-ULV.

It was of interest to determine if the 0.04 Pa offset observed between the two experiments during and immediately following the helium release was associated with the slightly different measured concentration distributions. The ΔP at the measurement location was calculated for 3600-LR-ULV immediately following the end of the helium flow using the same approach described earlier for 3600-LC-ULV. The calculated ΔP was -0.407 Pa as compared to the value of -0.395 calculated earlier. The change is in the correct direction, but it is only ¼ of the change observed experimentally. It must be concluded that the source of the small shift in pressure curves between 3600-LC-ULV and 3600-LR-ULV is not fully understood.

Despite the differences in concentration and differential pressure profiles between the two two-vent experiments with lower center and rear releases, the post release behaviors were quite similar. Comparisons of the values of $V\%_{t=7200}$ and $V\% \ slope_{t=7200}$ included in Table 3 for the various sensor heights show that the values were nearly identical. This indicates that the helium volume fraction measurements were accurate and that changes in response were not responsible of the small differences in concentration observed between the two experiments.

3.2.3. One Hour Helium Releases near the Ceiling at the Center of the Garage

The temporal concentration and differential pressure time profiles recorded during a one hour helium release from 2.5 cm below the center of the ceiling are shown in Figure 30 for the garage having a single 2.4 cm × 2.4 cm front-wall vent. Sensor #8 was located in the rear right-hand corner at the same height as sensor #1, (x,y,z,) = (141 cm, 134.5 cm, 9.3 cm). The air flow used to flush the garage was started at 26,399 s. Comparison with the concentration profiles for experiments with releases at the lower positions and the same size opening (see Figure 19 and Figure 23) shows that releasing helium near the ceiling led to much higher helium concentrations near the ceiling and a greater vertical stratification at the end of the release. These differences are evident when $V\%_{t=3600}$ values for the various sensor locations in Table 3 are compared with those for 3600-LC-SSV and 3600-LR-SSV. For the highest sensor location $V\%_{t=3600}$ was increased 23 %, and for the lowest it was decreased by 9 %. The result is that the average vertical concentration gradient was 120 % greater when helium was released near the ceiling of the garage. The vertical concentration gradients are non linear. For the release near the ceiling the gradients were largest in the upper portion of the enclosure, while the opposite was the case for releases near the floor.

Moving the helium release point from near the floor to near the ceiling also had a large effect on the time required for helium to initially reach the various sensor locations. As can be seen in Table 3, the values of t_{init} were nearly a factor of three longer in the upper region for the release near the ceiling and roughly a factor of two longer at lower measurement locations.

The observations concerning the effect of release height on helium distribution within the garage during the release phase suggest that much less mixing took place with the surrounding gas when the helium was released near the ceiling. This is consistent with the much shorter flow distance the released plume traveled before striking the ceiling. As a result, helium concentrations were higher, which in turn created a more stable upper layer that mixed more slowly with the gas below.

The pressure plot for 3600-UC-SSV in Figure 30 is subtly different than those for 3600-LC-SSV and 3600-LR-SSV discussed earlier. The differential pressures for all three cases showed a sharp increase when the helium flows first started. The differential pressure then fell roughly linearly until the helium

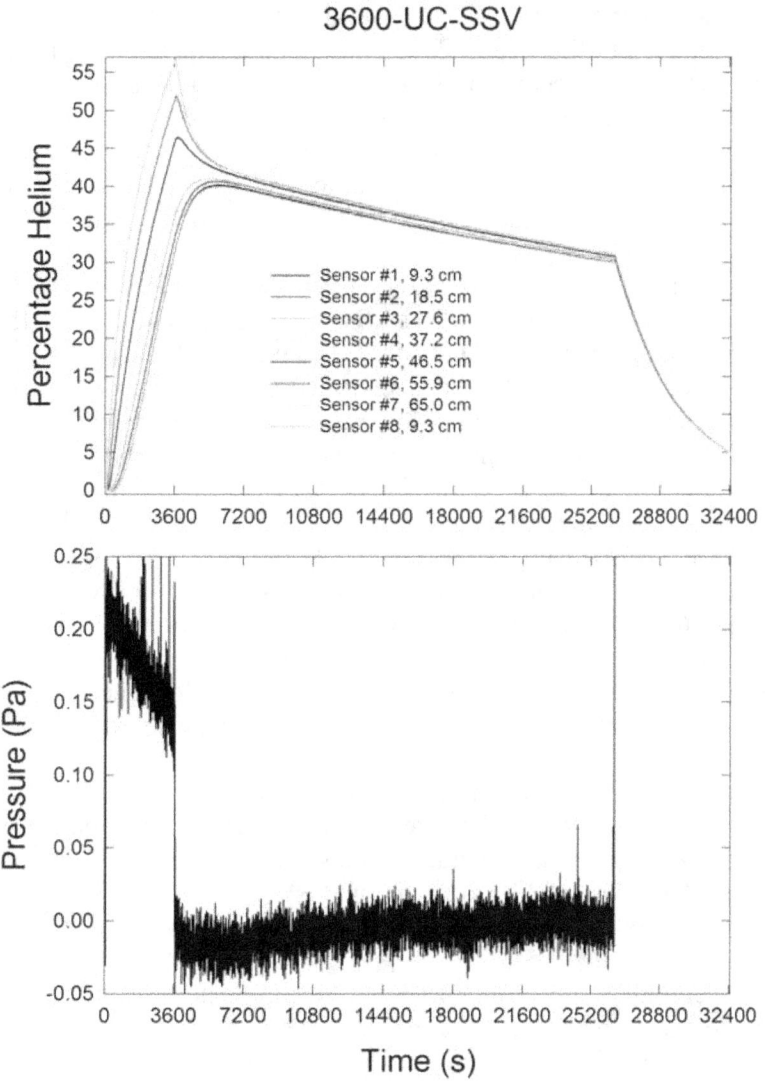

Figure 30. The upper plot shows helium volume percent as a function of time at eight locations for a one hour release of helium from the upper center position into the ¼-scale garage with a single centered 2.4 cm × 2.4 cm opening in the front face. The lower plot shows the differential pressure across the face.

flow was stopped. When the flow was halted, the differential pressure fell to values just above zero for the releases near the floor before slowly decaying towards zero. When the flow was halted for 3600-UC-SSV the differential pressure dropped below zero before beginning to recover towards zero. Despite these minor differences in differential pressure behavior, the ΔP values listed in Table 4 for the start and end of the helium release are in close agreement with those for the two releases near the floor.

Even though the mixing behavior inside the garage during the helium release was strongly affected by the release height, the same was not the case for the post-release period once the initial transient period has passed. As seen in Figure 30, when the helium flow was shut off the strong concentration gradients began to dissipate as the helium levels at the upper sensor locations dropped and those lower in the garage increased. Within 30 minutes only a mild concentration gradient remained. Similar behaviors are apparent in Figure 19 and Figure 27 for the releases near the floor. Values of $V\%_{max}$ and $t_{V\%max}$ in Table 3 provide insights into the response of the concentrations to halting the helium flow. Earlier it was shown for the experiments with releases near the floor that the helium concentrations at sensors #4 to #7 began to drop

43

immediately after the flow was shut off, while those lower down continued to rise for periods that increased closer to the floor. At the lowest position nearly 2300 s was required to reach the maximum concentration. Somewhat different behaviors were observed for the case with helium release near the ceiling. At the three highest positions, sensors #5 to #7, the helium concentrations continued to increase for brief periods before starting to fall, while a considerably longer period (roughly 650 s) was required for sensor #4 to reach its maximum value. At the lower locations 600 s longer periods were required than for the helium releases near the floor. Due to the higher concentration gradients, the ratios of $V\%_{max}$ to $V\%_{t=3600}$ for measurement positions near the floor were much greater than when helium was released near the floor. As an example, the ratios at sensor #1 were 1.1 for 3600-LC-SSV and 1.26 for 3600-UC-SSV.

By 3600 seconds after the end of the release the volume percent change from the sensor #1 to sensor #7 locations was only 2.3 % for 3600-UC-SSV. The comparable changes for the releases near the floor were 1.6 % and 1.5 %. Comparison of the concentration values, $V\%_{t=7200}$, at this time shows that the helium volume percents were around 1.9 % higher for the release at the ceiling as compared to those near the floor. This small 5 % increase suggests that slightly more helium was present inside the garage at the end of the release near the ceiling. The results indicate that even though the concentration vertical gradients depend strongly on release height, there is only a weak dependence of the total amount of helium in the garage at the end of the release on this parameter.

Values for the rates of decrease in helium volume percents at 3600 seconds after the ends of the releases have similar magnitudes for releases near the floor and ceiling. These values for 3600-UC-SSV are very similar to those for the releases near the floor at the lowest sensor positions and become slightly larger with increasing height. At the sensor #7 location the slope is 18 % larger. This observations suggest that the larger concentrations present at the end of the helium release are still dissipating 3600 seconds after the flow was stopped.

The air flow used to sweep helium from the garage was initiated before the eight hour period following the helium release had expired. The effect of the air flow is evident in Figure 30, where the helium concentrations start to decrease more rapidly at the same time that the differential pressure increase due to the air flow appears. The value of ΔP associated with shutting off the air flow is included in Table 4 and is in excellent agreement with the corresponding values observed during experiments with the helium released near the floor.

The temporal concentration and pressure profiles recorded for a one hour helium release at the upper-center location in the garage with a single large vent are shown in Figure 31. Sensor #8 was located in the back corner, (x,y) = (141 cm, 134.5 cm) at a height of 9.3 cm. The concentration profiles that developed during the 3600 s helium release are very similar to those recorded with the wall having a single small vent (see Figure 30). Comparison of $V\%_{t=3600}$ values for 3600-UC-SSV and 3600-UC-SLV in Table 3 shows that measured helium volume percents at the end of the helium release periods were slightly smaller (0.2 % to 0.4 %) for the larger vent. Recall that the corresponding values for helium releases near the floor at the center of the garage had differences of similar magnitude, but opposite sign, while those for rear releases near the floor were nearly identical. The changes are so small that it is impossible to identity a definite trend beyond the conclusion that the size of the centered vents had minimal effects on helium concentration profiles during the release phase.

Earlier it was shown that helium concentrations decayed measurably faster when the larger vent was used. The same is true for helium releases at the upper-center location. The quantitative measurements in Table 3 show that for the lower measurement locations the values of $V\%_{max}$ and periods required, $t_{V\%max}$ to reach maxima following the end of the helium release were reduced for 3600-UC-SLV as compared to 3600-UC-SSV. Similar reductions are evident at each of the vertical locations for helium concentrations measured 3600 s after the end of the release. As observed earlier, helium concentration fall off rates 3600 s after the end of the helium releases, $V\%$ $Slope_{t=7200}$ are nearly factors of two higher when the larger vent is used despite the concentration being somewhat less at this time. The ratios decrease from roughly 1.95 to 1.75 on going from the lowest to highest measurement locations.

The vent size had an apparent effect on the differential pressures recorded during the upper helium releases with single vents. Comparison of the profiles in Figure 30 and Figure 31 shows that following the

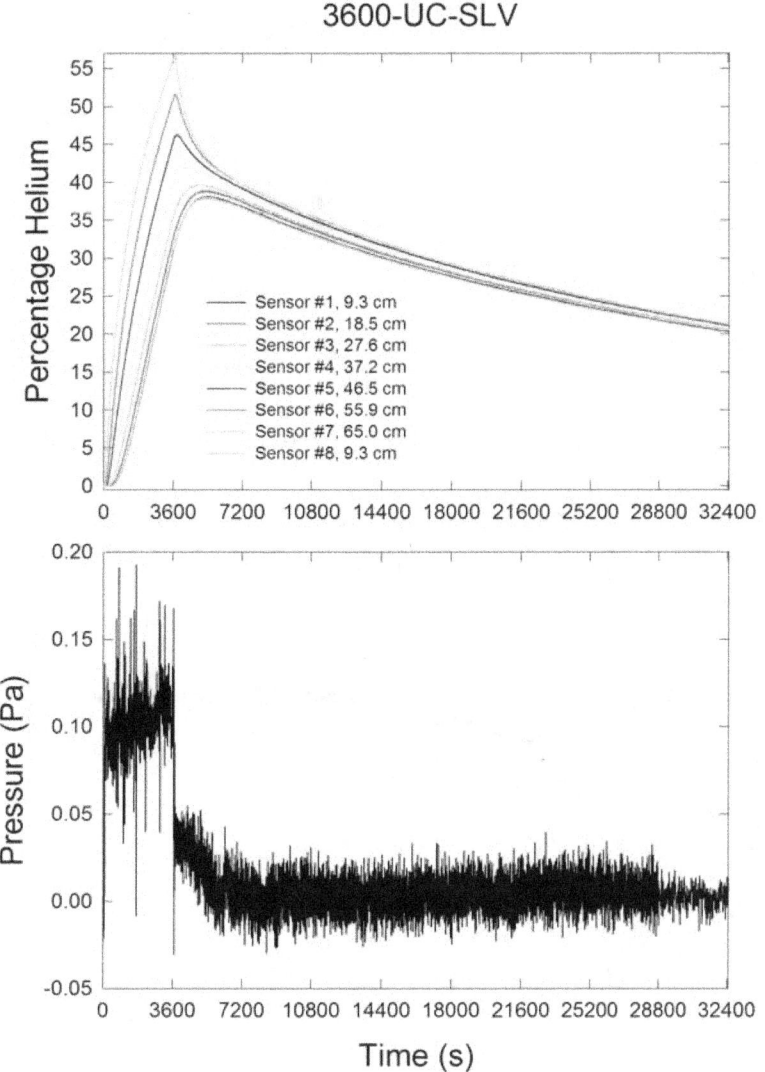

Figure 31. The upper plot shows helium volume percent as a function of time at eight locations for a one hour release of helium from the upper center position into the ¼-scale garage with a single centered 3.05 cm × 3.05 cm opening in the front face. The lower plot shows the differential pressure across the face.

initial rise due to the start of the helium flow, the differential pressure began to drop when the small vent was used and that it rose when the large vent was employed. When the helium flow was shut off the differential pressure dropped to slightly negative values with the small vent and to a positive pressure of higher magnitude with the larger vent. Similar differences were seen earlier in the shapes of the curves using the two vents for releases in the bottom of the garage. ΔP values at the start and end of the helium releases were very similar for 3600-LR-SLV and 3600-UC-SLV (Table 4).

The last one hour release case considered is a center release of helium near the ceiling with the two-vent face. Figure 32 shows the temporal concentration and differential pressure profiles. Sensor #8 was located just below the ceiling at $z = 73.5$ cm at a front location that mirrored the position of the vertical sensor array, i.e., (x,y) = (112.5 cm, 37.5 cm). The most notable feature of the concentration profiles is the strong helium concentration gradient that developed during the helium release. Comparison with the results with helium releases near the floor with two vents (see Figure 24 and Figure 29) shows that the vertical helium concentration profiles differed significantly. For the releases near floor the helium volume

45

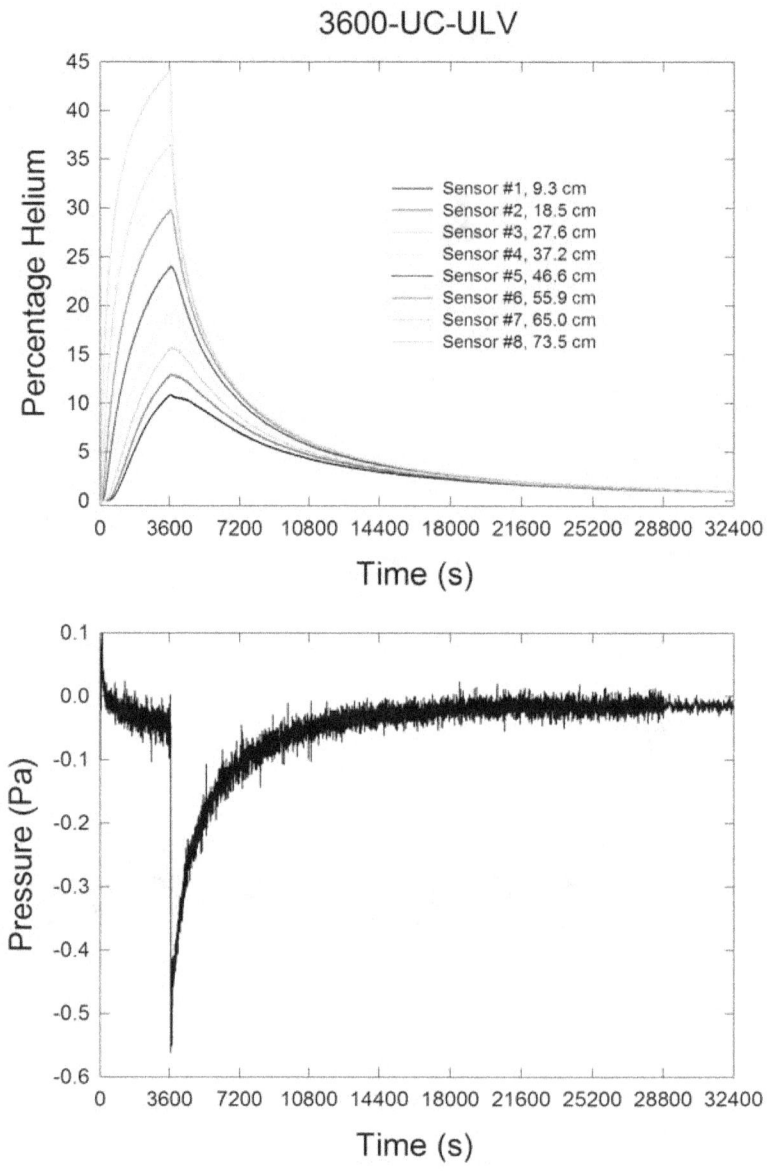

Figure 32. The upper plot shows helium volume percent as a function of time at eight locations for a one hour release of helium from the upper center position into the ¼-scale garage with 2.15 cm × 2.15 cm upper and lower openings in the front face. The lower plot shows the differential pressure across the face.

fractions were nearly constant in the upper layer and the gradient increased as the distance to the floor was reduced. The opposite was observed for the release near the ceiling, with the concentration and its gradient increasing continuously with height. The highest measured helium volume percent near the ceiling was 11 % higher for the upper release as compared to the nearly uniform upper-level concentration observed with the release near the floor. Interestingly, at the height of sensor #7 the concentrations were roughly equal, with values around 35 %. For lower measurement heights the concentrations for the upper release were smaller. At the 9.3 cm height the reduction in helium volume fraction was from roughly 17 % to 10 %. It is clear that release location had a large effect on the concentration distribution during the release phase.

The concentration profiles for 3600-LC-ULV and 3600-UC-ULV at the end of the releases are compared in Figure 33. Experimental measurements are indicated by solid symbols. Values at the floor and ceiling, estimated by visually extrapolating the experimental measurements to 0 cm and 75 cm,

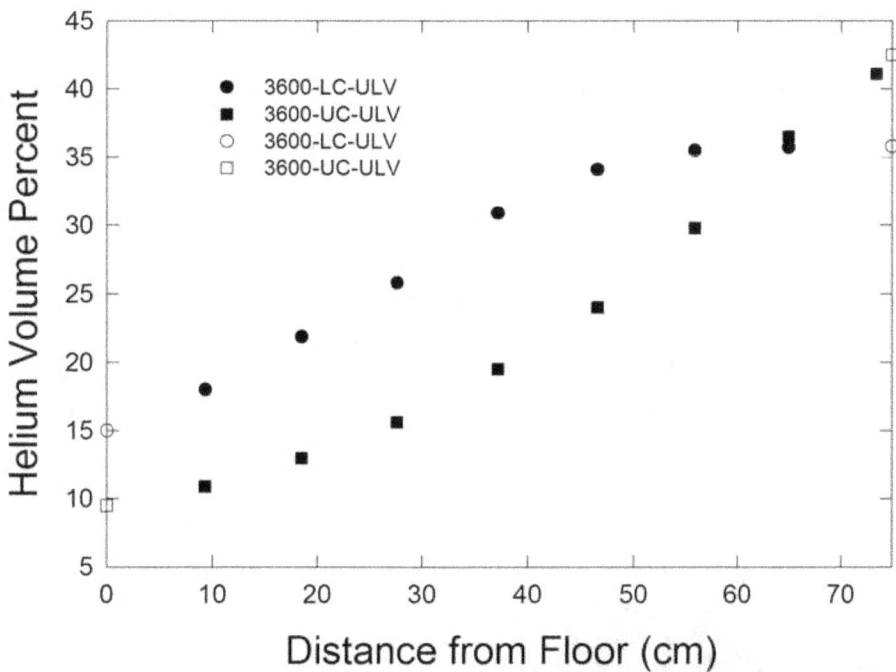

Figure 33. Values of experimental helium volume percents (solid symbols) are plotted as function of distance above the floor for 3600-LC-ULV (circle) and 3600-UC-ULV (square). Estimated values at the floor and ceiling are indicated by open symbols.

respectively, are indicated by open symbols. A simple vertical integration assuming that the piecewise steps are linear provided estimates for the average concentrations within the enclosure. The resulting helium volume fractions were 28.5 % and 22.1 % for the lower and upper releases, respectively. The results show that the amount of the helium in the garage at the end of the release was reduced by nearly 30 % for the upper release. Recall that only small changes in helium concentration profiles at the end of the releases were observed when upper and lower releases with single vents were compared.

Values of t_{init} included in Table 4 are similar to those observed for the upper release with a single vent even though there is some indication that longer periods were required to reach the lowest sensor positions.

When the helium flow was turned off, the helium concentrations began to fall almost immediately at each of the measurement locations, as shown by the values of $t_{V\%max}$ included in Table 3. The fall-off rates were much faster at the higher positions, and the relative differences in concentration with height were reduced. Comparison of the helium volume fractions measured 3600 s following the end of the helium flow in Table 3 shows that concentrations for 3600-UC-ULV were reduced 19 % to 23 % compared to those for 3600-LC-ULV. The relative differences are reduced somewhat from those at the end of the helium release, which is to be expected since helium loss rates from the garage were higher for higher concentrations. This is reflected in the values of $V\%\ slope_{t=7200}$ included in Table 4, which are 25 % to 30 % larger for 3600-LC-ULV.

The ΔP generated by starting the helium flow was small and comparable in magnitude to those observed when the single large-vent and two-vent faces were used in earlier experiments. Following the initial increase, the differential pressure decreased slowly, falling to -0.05 Pa by the end of the release. When the helium flow was shut off the differential pressure change was ΔP = -0.50 Pa. This value is slightly less than observed for the releases at the upper center location into the garage equipped with either

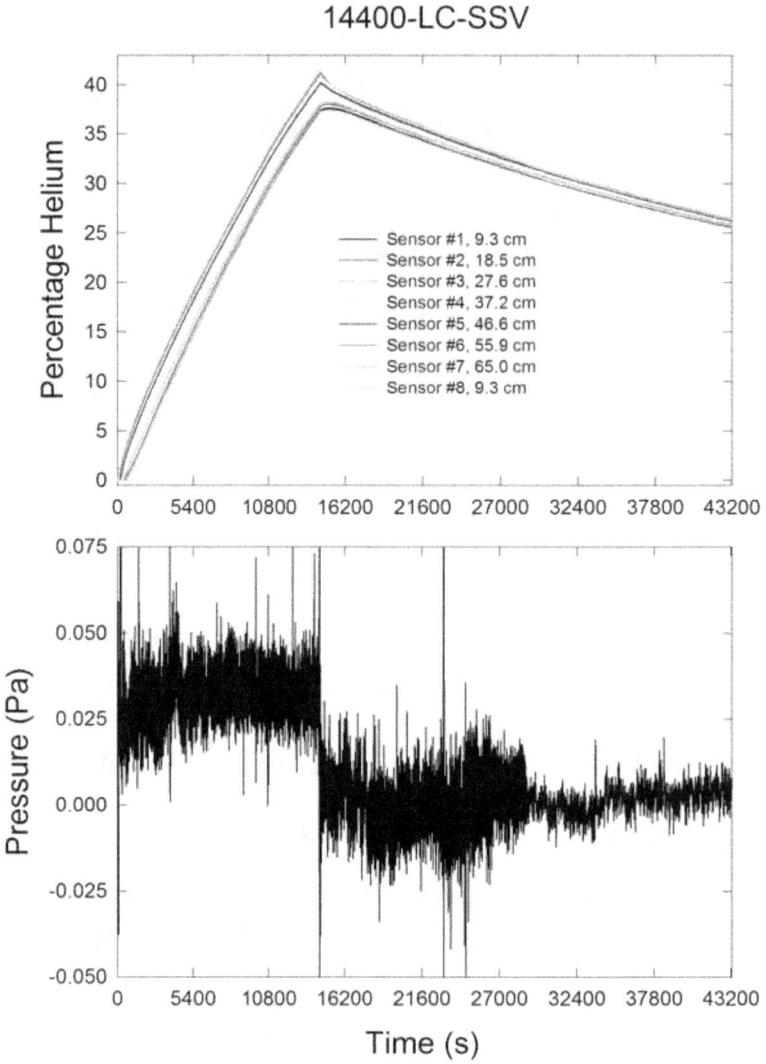

Figure 34. The upper plot shows helium volume percent as a function of time at eight locations for a four hour release of helium from the lower center position into the ¼-scale garage with a single centered 2.4 cm × 2.4 cm opening in the front face. The lower plot shows the differential pressure across the face.

of the two single vents. This is likely due to the different vertical concentration distribution when two vents were used.

3.2.4. Four Hour Helium Releases near the Floor at the Center of the Garage

Figure 34 shows concentration and differential pressure temporal profiles for a four hour release of helium into the garage equipped with a front wall having a single 2.4 cm × 2.4 cm vent. Sensor #8 was located at the same height as sensor #1 in the right rear corner, (x,y) = (141.0 cm, 134.5 cm). Comparisons with the corresponding case for a one hour release shown in Figure 19 and the quantitative values included in Table 3 show that the longer helium release led to concentrations at the end of the release period that were lower at the highest measurement position (sensor #7) and higher at the lowest measurement position (sensor #1), i.e., the concentration gradient was reduced with the slower release rate. In order to obtain an estimate for the average concentrations in the garage the same approach described above in which the

48

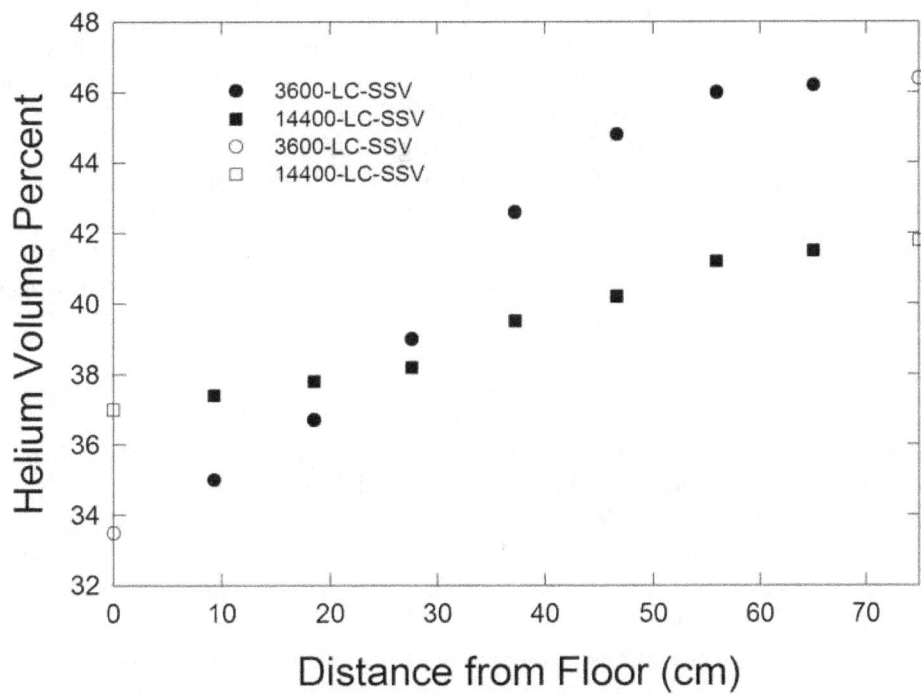

Figure 35. Values of experimental helium volume percents (solid symbols) are plotted as a function of distance above the floor for 3600-LC-SSV (circle) and 14400-LC-SSV (square). Estimated values at the floor and ceiling are indicated by open symbols.

vertical concentration profiles were integrated was used. Figure 35 shows plots of measured helium volume fractions at the end of the two releases at the seven measurement locations along with visually extrapolated values for $z = 0$ cm and $z = 75$ cm. The average values obtained from the integration are 41.3 % and 39.4 % for 3600-LC-SSV and 14400-LC-SSV, respectively. The average for the shorter release time is 5 % greater. However, based on the measured volume flow rates, the total volume of helium entering the garage was also 5 % larger. This indicates that there was no measureable difference between the amount of helium loss from the enclosure during the one hour and four hour releases.

As was to be expected, the lower flow volume rate did result in a longer filling time for the garage as compared to the one hour release. Values of t_{init} for the five highest sensor positions are roughly a factor of two larger for 14400-LC-SSV as compared to 3600-LC-SSV. The value at sensor #1 is roughly 80 % greater for the longer release, but the increased uncertainties in determining when the helium volume percents initially increase at the lower sensor locations should be kept in mind.

At the end of the release period the average vertical concentration gradient between the sensor #1 and sensor #7 locations was only 35 % as large as that at the end of the corresponding one hour release (helium volume fraction of 0.07 %/cm versus helium volume fraction of 0.20 %/cm). Similar to the one hour release, the concentrations at the upper sensor locations began to drop very shortly after the helium flow was stopped, while those at sensor #1 to sensor #3 continued to increase. However, since the initial gradients were smaller, the relative increases in concentration were smaller, and the times required to attain the maxima were reduced as can be verified by reviewing the results for $V\%_{max}$ and $t_{V\%max}$ for 14400-LC-SSV and 3600-LC-SSV in Table 3. For the one hour release the concentrations at sensors #1 to #3 continued to increase for 2100 s, 1500 s, and 1200 s after the flow was stopped, while the corresponding periods for the four hour release were 730 s, 500 s, and 260 s.

Table 3 includes values of helium volume percent, $V\%_{t=18000}$, and the temporal gradient, $V\% Slope_{t=18000}$, recorded one hour after the end of the four hour release at each sensor location. Comparison with the corresponding values of $V\%_{t=3600}$ shows that the measured helium volume fractions

49

for the four hour release were slightly lower (4 % to 5 %) than those for the one hour release. These differences are comparable to the difference in the amount of helium released between the two tests. These observations suggest that the loss rates of helium from the enclosure were proportionally the same. This is the expected result since the same single-vent face was used. The helium loss rates for the two tests had similar magnitudes, but the rates for 14400-LC-SSV were slightly higher at the lower sensor positions and slightly lower at the upper sensor positions. This may be an indication that the helium concentrations were still adjusting to the different vertical distributions that existed at the end of the helium flow.

The differential pressure plot in Figure 34 shows that the pressure changes associated with the four hour helium release were small and barely detectable due to the noise limit of the detector. A value of ΔP = 0.025 Pa was measured when the helium flow was initiated. This can be compared with ΔP = 0.22 Pa measured for 3600-LC-SSV. Using Eq. (2) and the value of n = 0.531 from Table 2, the predicted ratio for the ΔP values is 15.0, which is considerably higher than the observed ratio of 8.8. Earlier it was shown that Eq. (2) gave poorer agreement with experiment as ΔP became smaller. This is likely the reason for the poor agreement seen here. During the release the differential pressure remained roughly constant. When the helium was shut off the observed pressure change, ΔP, was slightly smaller as expected since a helium-air mixture was flowing through the vent. When air was used to flush the garage, the ΔP measured when the air flow was halted was in excellent agreement with other measurements using the same face.

Figure 36 shows the results of a four hour release at the center of the garage equipped with the single large vent. Comparison with Figure 35 shows that the helium volume fraction curves have a similar appearance during the helium release periods, but that the concentrations at a given height are slightly reduced with the larger opening. These reductions are evident by comparing values of $V\%_{t=14400}$ for the two conditions. By integrating the concentration profiles along the vertical direction in the manner described earlier, approximate average volume fractions for the garage were calculated. The results gave 39.4 % and 37.1 % for 14400-LC-SSV and 14400-LC-SLV, respectively. The vent size had an effect on the amount of helium loss from the garage for center helium releases near the floor for four hour releases. Recall that no vent-size effect was evident for one hour helium releases near the floor. Values of t_{init} for the eight sensors in Table 4 are very similar for the two releases except at the sensors #1 and #8 locations, where slightly longer periods were required for helium to reach these locations when the larger vent was used. This behavior is different than observed for the one hour releases with single vents for which values of t_{init} were substantially increased when the single vent size was increased.

Figure 37 compares the vertical concentration profiles at the ends of the helium releases for 3600-LC-SLV and 14400-LC-SLV. These curves were integrated to determine the approximate average helium volume fractions for the garage. The results were 41.6 % and 37.1 %. The result for the one hour release was 12 % higher. Since the total helium flow volume was only 5 % higher for the one hour release, this result provides additional evidence that more helium was loss from the garage during a four hour release near the floor with the large vent in place.

The differences in the amounts of helium in the garage at the end of the releases carried over into the post-release loss of helium from the enclosure. Comparisons of the helium volume fractions one hour after the end of the releases included in Table 3 show that the concentrations were considerably reduced at all sensor heights for 14400-LC-SLV as compared to the one hour releases near the floor with single vents and to the center four hour release near the floor with the single small vent. The effect of the increased loss of helium during the release is also reflected in the rates of helium loss at t = 18000 s. Results discussed earlier showed that helium loss rates were dependent on the opening size, but did not vary significantly for runs with the same face since the average amount of helium present at the end of a release did not depend on vent size. This is not the case for 14400-LC-SLV, for which more helium was loss during the release. Comparison of $V\%\ Slope_{t=18000}$ values for the various sensors indicates that helium loss rates for 14400-LC-SLV are considerably reduced as compared to those for 3600-LC-SLV. This is likely due to the fact that the helium loss rate from the enclosure decreased as the helium levels fell.

Due to a technical problem, the initial 41 minutes of the differential pressure curve shown in Figure 36 was not recorded. When the pressure measurements became available during the helium release, a very small negative differential pressure was evident. The ΔP associated with halting the helium flow was on

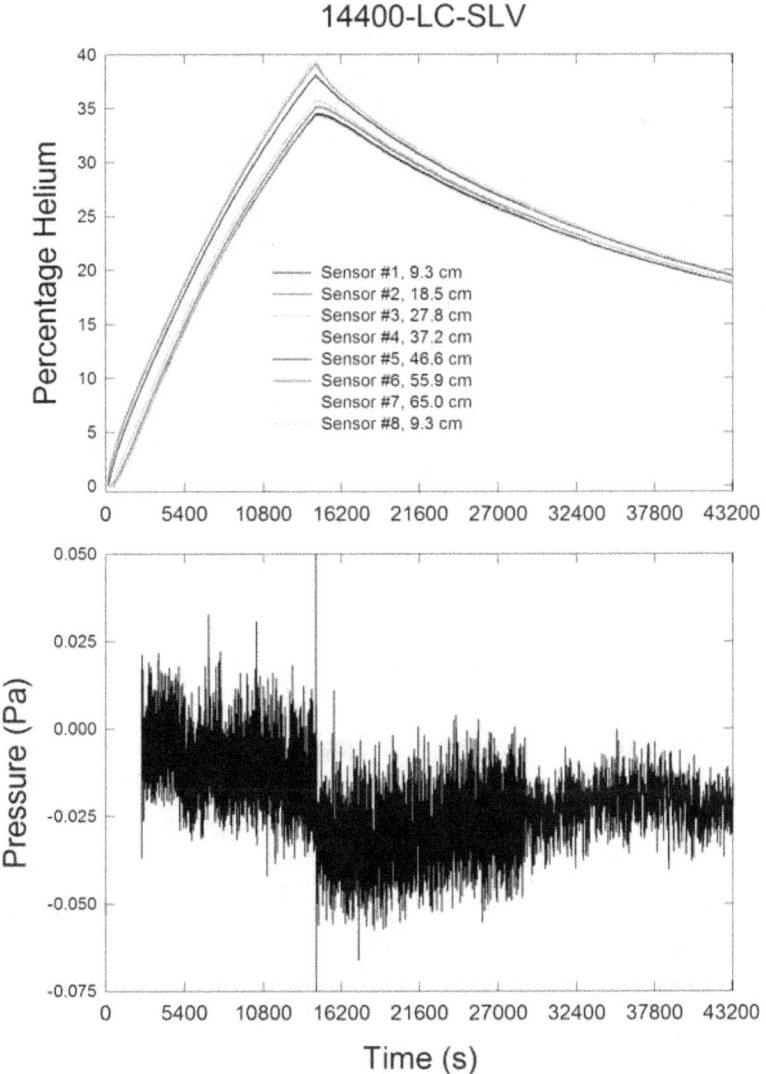

Figure 36. The upper plot shows helium volume percent as a function of time at eight locations for a four hour release of helium from the lower center position into the ¼-scale garage with a single centered 3.05 cm × 3.05 cm opening in the front face. The lower plot shows the differential pressure across the face.

the order of − 0.025 Pa. The ΔP observed when the air flow was halted after sweeping out the remaining helium was 0.362 Pa. This value agrees well with that for 3600-LR-SLV included in Table 4.

The concentration and differential pressure profiles for a four hour release into the garage with the front face having 2.15 cm × 2.15 cm upper and lower vents are shown in Figure 38. Sensor #8 was located above the release point 65.0 cm from the floor. Two properties of the time behaviors during the helium release stand out. First, the helium concentrations at the end of the release are much lower than observed for the experiments discussed thus far, and second, the concentrations are approaching a steady state after four hours, with the profiles having begun to flatten out starting around three hours. These observations indicate that not only was significant helium flowing out of the garage, but that large amounts of air were also flowing in during the release. Recall that this was not the case when single vents were used, with the possible exception of 14400-LC-SLV.

Comparison of $V\%_{t=14400}$ values at the sensor #7 location confirms that for this upper sensor location the helium volume fraction at the end of the helium release was more than a factor of two smaller

51

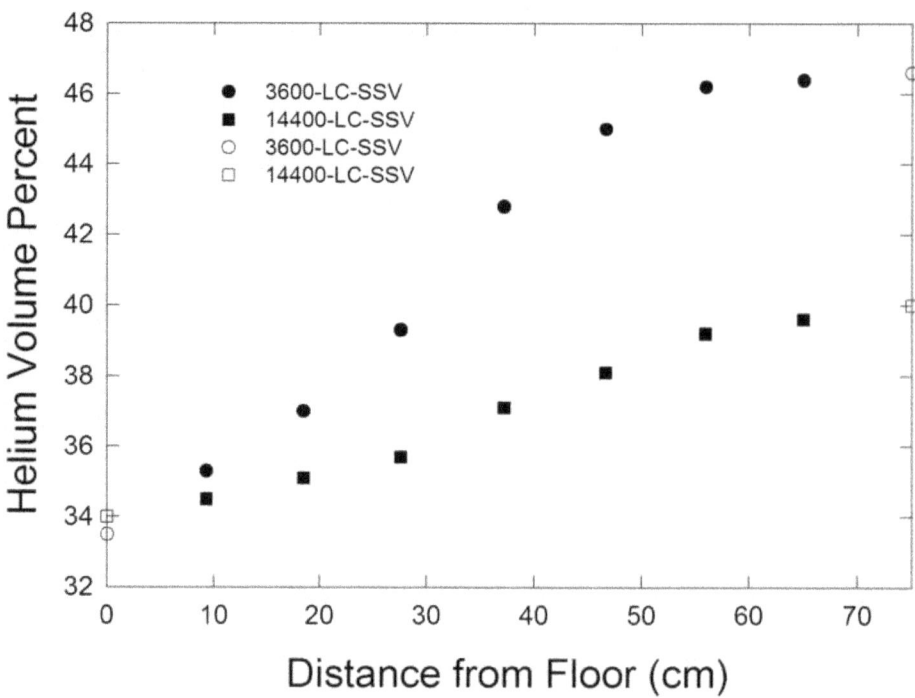

Figure 37. Values of experimental helium volume percents (solid symbols) are plotted as a function of distance above the floor for 3600-LC-SLV (circle) and 14400-LC-SLV (square). Estimated values at the floor and ceiling are indicated by open symbols.

than recorded for the other tests discussed thus far. This includes the comparable one hour release with two vents. Interestingly, the periods required for the helium to first reach the sensor locations were nearly identical to those for 14400-LC-SLV, which had a comparable vent area. This indicates that the initial filling behavior was not strongly influenced by the vent configuration. This could have been expected since helium concentrations were initially low, and hydrostatic pressure variations large enough to drive flows into and out of the garage had not yet developed.

As observed for the comparable one hour release, the measurements directly above the release point indicated that significant mixing of the helium plume with its surroundings took place between the lower release point and the sensor height of 65.0 cm. The one second averaged concentrations also have considerably higher fluctuations inside the buoyant flow than were observed outside of the plume.

Either immediately or shortly after the helium flow was shut off after 14400 s, the helium concentrations at all seven of the sensor locations began to fall. After this, there was a rapid decrease in helium concentration, and the vertical concentration gradient slowly decreased. Within an hour, the concentrations inside the garage dropped by more than a factor of two in the upper part of the garage and slightly less than a factor of two in the lower part as confirmed by comparing $V\%_{t=14400}$ and $V\%_{t=18000}$ values in Table 3. The helium volume percent fall off rates for 14400-LC-ULV had lower values compared to those for the corresponding one hour releases with two vents. This difference tracks the corresponding ratios of helium volume fraction at the ends of the helium releases. The exchange rates at these times were dominated by the hydrostatic pressure differences, and these were smaller due to the lower helium concentrations that developed during the 4-hour release.

The differential pressure behavior shown in Figure 38 also reflects the dominance of hydrostatically-induced pressure flows for this case. When the helium flow was initiated, the ΔP was just barely detectable in the noise. The differential pressure dropped slowly to a value of roughly -0.055 Pa by the end of the helium release. When the helium flow was shut off, the pressure dropped abruptly by $\Delta P =$ -0.085 Pa to -0.14 Pa. As can be seen in Table 4, this ΔP was more than a factor of five smaller than those

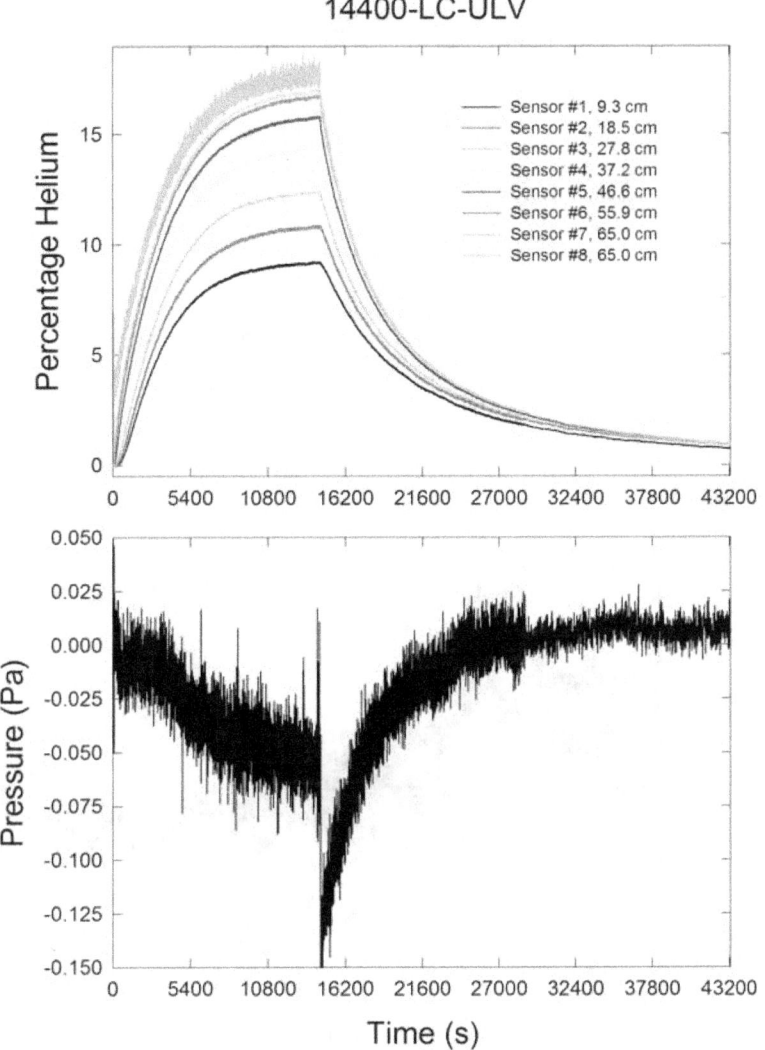

Figure 38. The upper plot shows helium volume percent as a function of time at eight locations for a four hour release of helium from the lower center position into the ¼-scale garage with 2.15 cm × 2.15 cm upper and lower openings in the front face. The lower plot shows the differential pressure across the face.

observed for the one hour flows with the two-vent face. The vertical concentration profile was used to calculate density and hydrostatic pressure profiles in the manner described earlier. Using the same assumptions, the required differential pressure ratio between the top and bottom vents was calculated to be 0.853. When the differential pressure was adjusted to match this pressure ratio, the value at $z = 37.5$ cm was -0.132 Pa. This is close to the measured value of -0.14 Pa and confirms the dominance of hydrostatic pressure differences in the observed differential pressure measurements and in the observed high helium loss rates for the experiments with two vents. When the garage was flushed the air, the measured ΔP was -0.323 Pa when the flow was shut off. This value agrees well with others in Table 4 for experiments with the two-vent face.

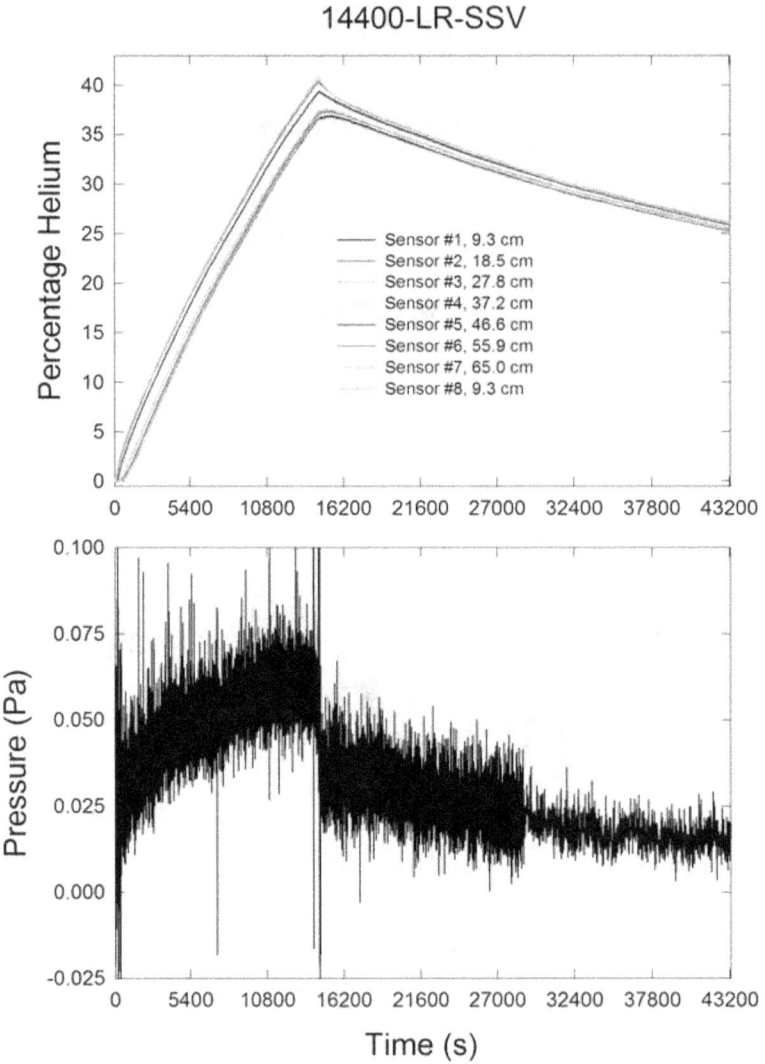

Figure 39. The upper plot shows helium volume percent as a function of time at eight locations for a four hour release of helium from the lower rear position into the ¼-scale garage with a single centered 2.4 cm × 2.4 cm opening in the front face. The lower plot shows the differential pressure across the face.

3.2.5. Four Hour Helium Releases near the Floor at the Rear of the Garage

Figure 39 shows the concentration and differential pressure time profiles for the release point centered at the lower rear position and the garage equipped with the single small vent. Sensor #8 was placed 9.3 cm above the floor at the same lower right-rear location as for the other single-vent cases. The concentration profiles for the eight sensor locations appear similar to those recorded for 14400-LC-SSV (see Figure 34). However, comparison of the $V\%_{t=14400}$ values included in Table 3 shows that all of the concentrations at the end of the rear release had helium volume fractions which were about 0.008 less than the corresponding values for the center release. This difference between front and rear releases is larger than observed for any of the other cases where center and rear lower releases were compared for the same release time and vent sizes. No obvious reason for the difference has been identified. It remains unclear if

54

the shift in concentration profiles was due to moving the helium release point or to some variation in the test system or condition.

The values of t_{init} included in Table 4 for 14400-LR-SSV were similar to those observed for 14400-LC-SSV. The differences that appeared in the concentration profiles were apparently not manifested in the initial mixing behavior.

The concentration responses immediately following the end of the helium release were consistent with those already described for cases with single vents. The concentrations at sensors #4 to #7 either began to fall immediately or very shortly afterwards. Helium volume fractions continued to rise at the lower sensor locations, with the time required to reach a maximum increasing as the height decreased. Each $t_{V\%max}$ at the lower positions was slightly larger for 14400-LC-SSV as compared to 14400-LR-SSV, reflecting the slight difference in concentrations at the end of the release period.

The difference in $V\%_{t=14400}$ between the center and rear releases carried over into the concentrations present one hour after the helium flow stopped. The helium volume fractions for each sensor location were either 0.006 or 0.007 lower for the rear-release position. On the other hand, the helium volume percent fall off rates were very similar for these two flows, which indicates that the small observed concentration differences did not significantly affect the rate of exchange of the inside helium-air mixture and the outside air.

The differential pressure changes in Figure 39 are very small. Following a 0.025 Pa increase as the helium flow started (see Table 4), the differential pressure increased slightly throughout the helium release period. When the helium was shut off at $t = 14400$ s the differential pressure fell by 0.020 Pa to a slightly positive value. The differential pressure curves for the center release in Figure 34 did not rise as much during the release phase, but the ΔP values associated with the beginning and end of the helium flow were of comparable magnitudes since the same vent size and helium volume flow rate were used. The $\Delta P = -0.363$ Pa measured when the air flow was shut off is comparable to other measurements with the single small vent.

The experimental results with the single 3.05 cm × 3.05 cm vent for the four hour rear release are shown in Figure 40. Once again sensor #8 was placed near the rear wall at the same height as sensor #1. The concentration profiles appear very similar to those in Figure 36 for the corresponding release at the lower center position. Comparison of $V\%_{t=14400}$ values for the eight sensor locations shows that the values for 14400-LC-SLV and 14400-LR-SLV were nearly identical. The differences in helium volume fraction for sensors #6 and #7 were -0.002 and -0.03. For the other sensors the differences were either -0.001 or zero. The close agreement between these two experiments suggests that the somewhat larger differences observed between 14400-LC-SSV and 14400-LR-SSV were the result of some systematic difference in the experiment and not the result of moving the release location from the center to rear location. As observed for the center release, the helium concentrations at the end of the lower-rear helium release were somewhat reduced when the larger vent was used, suggesting that some air was exchanged with the outside during the release period.

The values of t_{init} in Table 4 show that the times of initial helium detection were similar to those observed for the comparable center release, with some scatter. As for other similar comparisons, the periods required for initial helium detection at a given sensor were considerably longer than observed for 14400-LR-SSV.

The post-release concentration behaviors are consistent with earlier observations. Immediately following the end of the release the helium concentrations at sensors #4 to #7 began to drop, while those at the lower locations continued to rise for short periods. Both $V\%_{max}$ and $t_{V\%max}$ at a given sensor were reduced as compared to 14400-LR-SSV. However, the values were similar to those for 14400-LC-SLV, indicating a minimal effect of release location on post-release mixing.

The differential pressure increase when the helium flow was halted is barely discernible at $\Delta P = 0.005$ Pa. A similar value was observed during 14400-LC-ULV, which had nearly the same total vent area. When the helium was shut off, the differential pressure dropped 0.02 Pa and then rose slowly back to zero as the helium concentration inside the garage dropped.

55

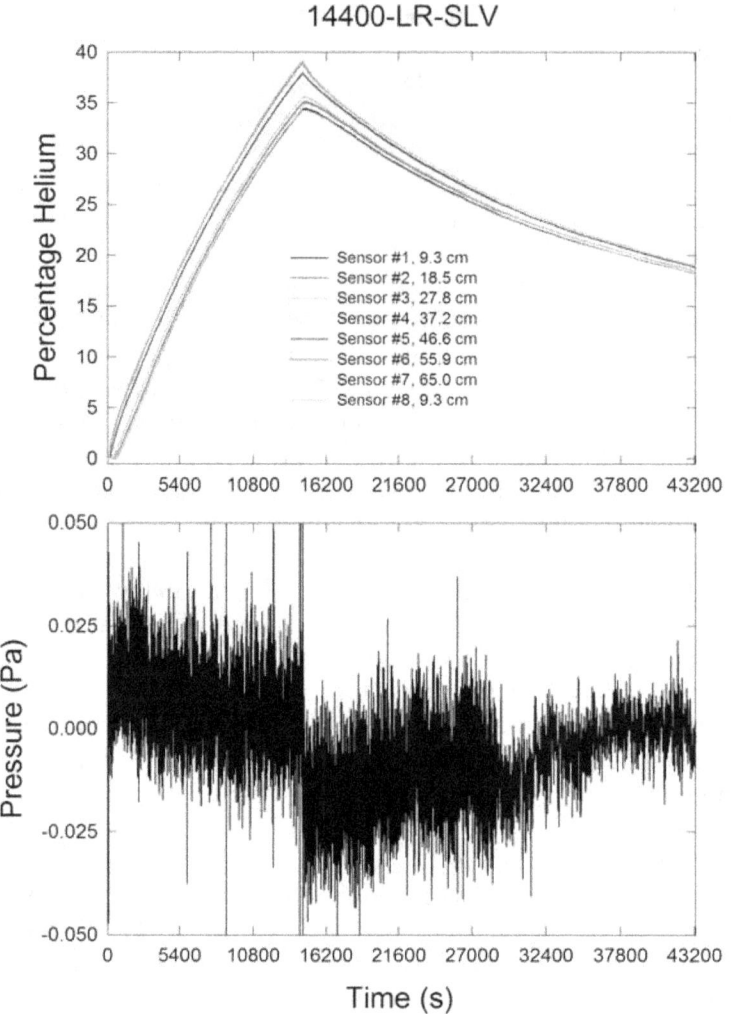

Figure 40. The upper plot shows helium volume percent as a function of time at eight locations for a four hour release of helium from the lower rear position into the ¼-scale garage with a single centered 3.05 cm × 3.05 cm opening in the front face. The lower plot shows the differential pressure across the face.

The results for a rear-release near the floor into the garage equipped with the two-vent front wall are shown in Figure 41. The concentration profiles for the eight sensor locations were very similar to those for 14400-LC-ULV. A sensor-by-sensor comparison of $V\%_{t=14400}$ values shows that measured helium volume fractions at the end of the release were slightly lower (0.002 or 0.003) for the release in the rear. As discussed earlier, it is not possible to determine if this was due to experimental uncertainty or the change in helium release location. Values of t_{init} listed in Table 3 for the eight sensor locations show some variations, but fall close to each other. Once again, the location of the release point in the lower part of the garage had little or no effect on the development of the helium concentration field during the helium release.

At the one hour point following the end of the helium release, the helium concentrations had fallen off by roughly one half from their maximum values. Values of $V\%_{t=18000}$ still had a significant vertical gradient. Quantitative comparison with the corresponding results for 14400-LC-ULV indicates that nearly all of the helium volume fractions agreed within 0.001, with the rear-release results being lower. The differences are reduced somewhat compared to $V\%_{t=14400}$. Interestingly, the fall off rates included in

56

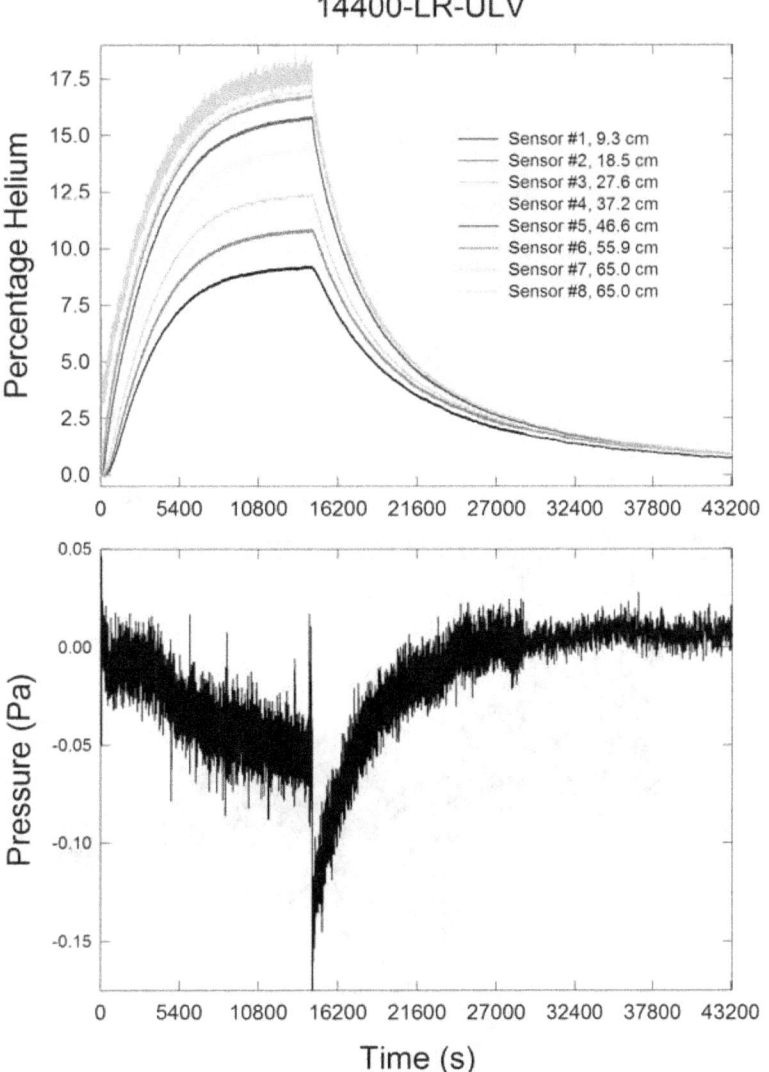

14400-LR-ULV

Figure 41. The upper plot shows helium volume percent as a function of time at eight locations for a four hour release of helium from the lower rear position into the ¼-scale garage with 2.15 cm × 2.15 cm upper and lower openings in the front face. The lower plot shows the differential pressure across the face.

Table 3 were slightly lower for the rear-release location, which provides an explanation for why the values of $V\%_{t=14400}$ moved closer together.

The differential pressure behavior for the rear release was very similar to that for the center release shown in Figure 38. Following a roughly $\Delta P = 0.1$ Pa (see Table 4) increase when the helium flow was initiated, the differential pressure decreased roughly -0.06 Pa by the end of the release period. When the helium flow was stopped, the differential pressure dropped lower by $\Delta P = -0.09$ Pa. These behaviors are in good quantitative agreement with those seen for 14400-LC-ULV. The value of $\Delta P = -0.316$ Pa measured when the air flow was turned off agrees well with similar measurement in Table 4 when this face was used.

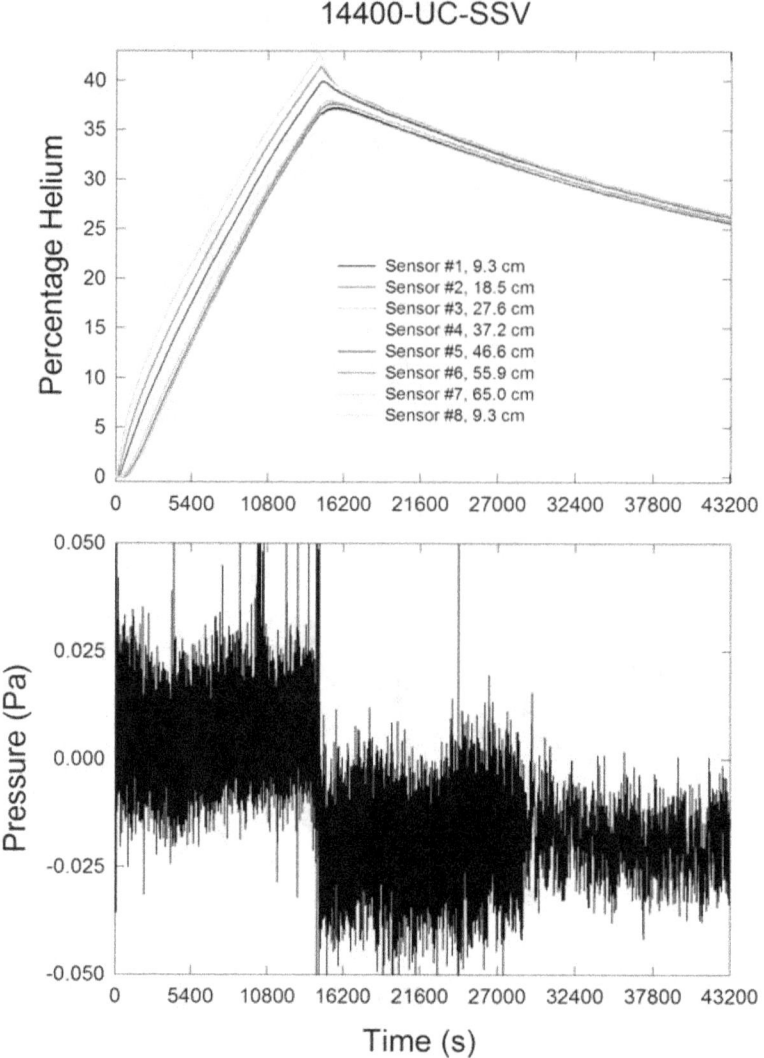

Figure 42. The upper plot shows helium volume percent as a function of time at eight locations for a four hour release of helium from the upper center position into the ¼-scale garage with a single centered 2.4 cm × 2.4 cm opening in the front face. The lower plot shows the differential pressure across the face.

3.2.6. Four Hour Helium Releases near the Ceiling at the Center of the Garage

The helium concentration and differential pressure time profiles for the garage with the single small vent are shown in Figure 42 for a four hour helium release near the ceiling. Sensor #8 was located at the 9.3 cm height near the rear wall on the right side. The maximum helium volume fractions at the end of the helium release were of the same order of magnitude as for the releases near the floor (compare with Figure 34 and Figure 39) with the same face. Close inspection reveals that the average concentration gradient is slightly higher for the release near the ceiling and that the locations of the largest concentration gradients (near the ceiling for 14400-UC-SSV and near the floor for 14400-LC-SSV and 14400-LR-SSV) are reversed. The differences are confirmed by the quantitative values listed in Table 3. These observations are similar to those for the one hour releases, but the gradients were considerably reduced. For the one hour release the average vertical gradient between sensors #1 and #7 at the end of the helium

58

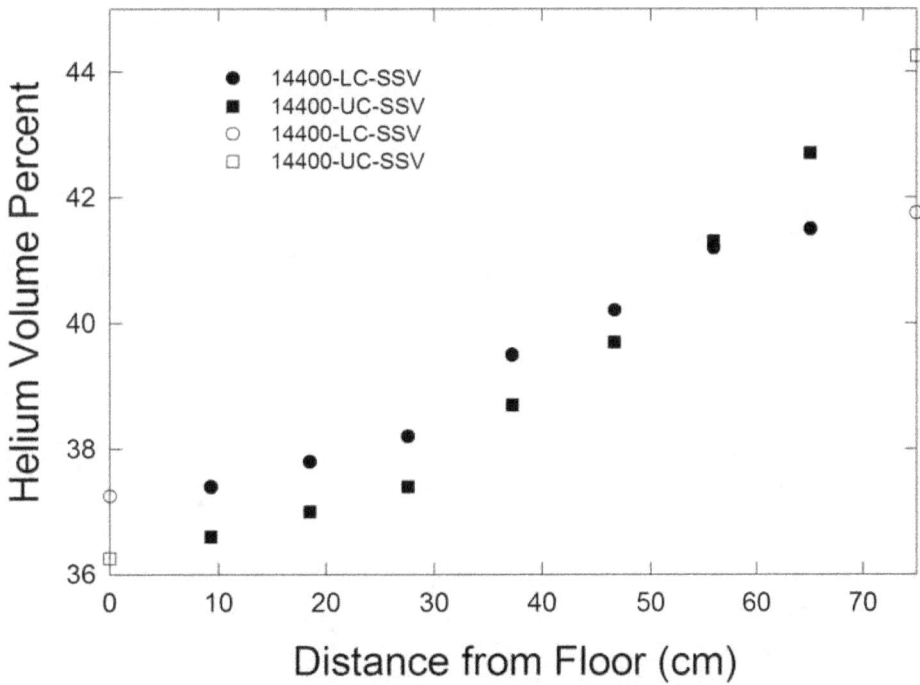

Figure 43. Values of experimental helium volume percents (solid symbols) are plotted as a function of distance above the floor for 14400-LC-SSV (circle) and 14400-UC-SSV (square). Estimated values at the floor and ceiling are indicated by open symbols.

release increased by 120 % when the release location was moved from near the floor to near the ceiling. The increase for the four hour release was only 40 %.

Values of $V\%_{t=14400}$ for the lower-center and upper-center releases are plotted in Figure 43 along with extrapolated values at the ceiling and floor. When these sets of data were integrated, average helium volume fractions of 0.394 and 0.392 were calculated for 14400-LC-SSV and 14400-UC-SSV, respectively. These results show that even though the release location had an effect on the concentration distribution at the end of the release period, the total amount of helium inside the garage was independent of this parameter when the single small vent was used.

Earlier comparisons showed that moving the helium release location from near the floor to near the ceiling for one hour releases substantially increased the period required for helium to reach the various sensor heights. The data in Table 3 indicates similar increases in t_{init} for the four hour releases with the small vent. Increases near the ceiling were on the order of a factor of three while those closer to the floor were on the order of two.

Similar to the observations for the corresponding one hour helium release near the ceiling, helium concentrations at all of the sensor locations continued to increase during the early portion of the post-release period. As indicated by the values of $t_{V\%max}$ in Table 3, the increases were of fairly short duration near the ceiling and somewhat longer closer to the floor. Comparison of $t_{V\%max}$ values shows that the periods following the end of a release were somewhat shorter near the floor and somewhat longer near the ceiling for the four hour release as compared to the one hour release.

Due to the differences in concentration distributions, the redistribution of the concentration profile following the end of the helium release was larger for the upper helium release location as compared to those at the lower positions, but the concentrations for all of the sensor locations approached a common concentration. By one hour following the end of the release, helium concentrations, $V\%_{t=18000}$, were nearly identical for 14400-UC-SSV and 14400-LC-SSV as seen in Table 3. Comparison of the fall-off rates at these times shows that those for 14400-UC-SSV were slower by about 10 %.

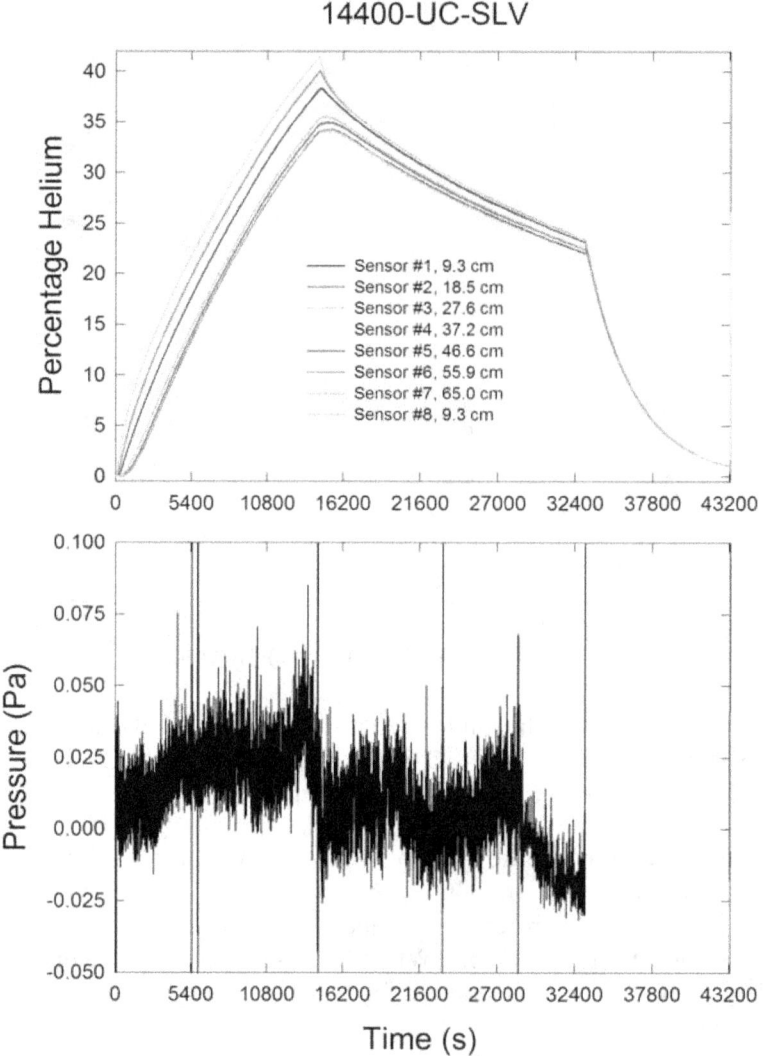

Figure 44. The upper plot shows helium volume percent as a function of time at eight locations for a four hour release of helium from the upper center position into the ¼-scale garage with a single centered 3.05 cm × 3.05 cm opening in the front face. The lower plot shows the differential pressure across the face.

The differential pressure plot was too noisy to determine a value of ΔP when the helium flow was initiated. However, as evident in Figure 42, the pressure drop when the helium flow was stopped could be estimated as $\Delta P = -0.025$ Pa, which is close to the corresponding value for 14400-LC-SSV. The $\Delta P = -0.987$ Pa associated with halting the air flow was in good agreement with others recorded for the garage with a single small vent.

Helium concentration and differential pressure measurements with the front wall containing the single large vent confirm many of the observations discussed earlier. The data are plotted in Figure 44, and the numerical values are included in Table 3. Sensor #8 was located at the same location in the lower-right rear as for the other single vent experiments. Note that the air flow used to flush the garage was started at 33,110 s for this experiment. The helium volume fraction curves have similar appearances to those for the four hour release near the ceiling with the small vent. However, the maximum volume fractions were reduced somewhat at all sensor heights, indicating that more air infiltrated into the garage from outside when the larger vent was used. This agrees with the conclusion from four hour releases near the floor with the same front walls. The vertical concentration profiles differ somewhat from those for which helium was released near the floor, with higher concentrations near the ceiling and somewhat lower values near the floor. This agrees with similar comparisons for both one hour and four hour releases. The vertical concentration profiles for 14400-LC-SLV and 14400-UC-SLV were integrated to obtain spatially-averaged helium volume fractions of 0.370 and 0.374. It is unclear whether the small difference is significant or not, but it is safe to say that the helium release height did not have a large effect on the average concentrations of helium present in the enclosure at the end of the releases.

The post-release concentration behavior was also similar to that observed earlier. Consistent with the data for 14400-UC-SSV, following the shut off of the helium flow, the concentrations at all of the sensor locations continued to rise. At the highest sensor locations, the increases only took place for short periods before the concentrations began to decrease, while longer periods were required closer to the floor. In general, the times to reach the concentration maxima, $t_{V\%max}$, were shortened significantly as compared to 14400-UC-SSV.

Helium volume fractions one hour after the end of the helium release were considerably reduced for the release with the larger vent in the front face as compared to those with the smaller vent. On the other hand, the helium volume fractions at all of the sensor locations were only slightly higher (generally by 0.003) for 14400-UC-SLV as compared to 14400-LC-SLV. The small differences are similar to those seen in the concentration profiles at the ends of the helium releases for these two cases. Values of $V\%\ Slope_{t=18000}$ listed in Table 3 are roughly 40 % higher for 14400-UC-SLV as compared to 14400-UC-SSV. When the values for 14400-UC-SLV are compared with those for 14400-LC-SLV, the former are found to be slightly larger (order of 7 %) for all of the sensor locations. These differences may reflect the slightly higher helium volume fractions observed for the release near the ceiling or may be due to small systematic differences similar to those that have shown up in other such comparisons.

The differential pressure plot included in Figure 44 shows that changes were very small with the large single vent. For some reason, there was less noise in these measurements as compared to 14400-UC-SSV, and it was possible to estimate ΔP as 0.01 Pa when the helium flow was started. As evident in the figure, the differential pressure rose slightly during the release, reaching a value of roughly 0.04 Pa by the time the flow was halted. At his point, the pressure dropped by ΔP = -0.2 Pa. Afterwards, the differential pressure dropped towards zero as the helium concentration inside the garage fell. The pressure drop that resulted from halting the air flow used to sweep helium from the garage was ΔP = -0.373 Pa, which is very close to other measurements with the single large vent.

The last experiment considered is a four hour helium release from the upper-center location into the garage equipped with upper and lower vents. Figure 45 shows the helium volume fraction and differential pressure time profiles. Sensor #8 was located at a height of 73.5 cm at the center of the right-front quadrant of the garage. The concentration profiles have shapes similar to those observed for the four hour releases at the lower locations with the same front wall. The helium volume fractions initially increased rapidly during the first hour and a half of the release period and then began to level off. At the end of the helium release the concentrations were still increasing slowly. The values of $V\%_{t=14400}$ were somewhat lower than those observed for the four hour releases at lower locations with the same wall over most of the height of the garage. Figure 46 compares $V\%_{t=14400}$ values as a function of height for 14400-LC-ULV and 14400-UC-ULV. Note that the value for sensor #8 for 14400-UC-ULV, which was located just below the ceiling on the opposite side of the garage from the vertical sensor array, for the upper release is consistent with the extrapolated value at the ceiling. The vertical profiles were integrated in the way described earlier and average helium volume fractions of 13.6 % and 11.8 % were calculated for the lower and upper release locations, respectively. Roughly 13 % more helium was lost from the enclosure during the release near the ceiling. This is consistent with the higher helium volume percent at this height for the release near the ceiling.

Figure 33 shows a plot of $V\%_{t=14400}$ versus height for the comparable experiments using a one hour release of helium. While the helium concentrations were significantly higher for the one hour release, the relative vertical variations had similar shapes. For the one hour release it was estimated that nearly 30 % more helium was lost during the release near the ceiling.

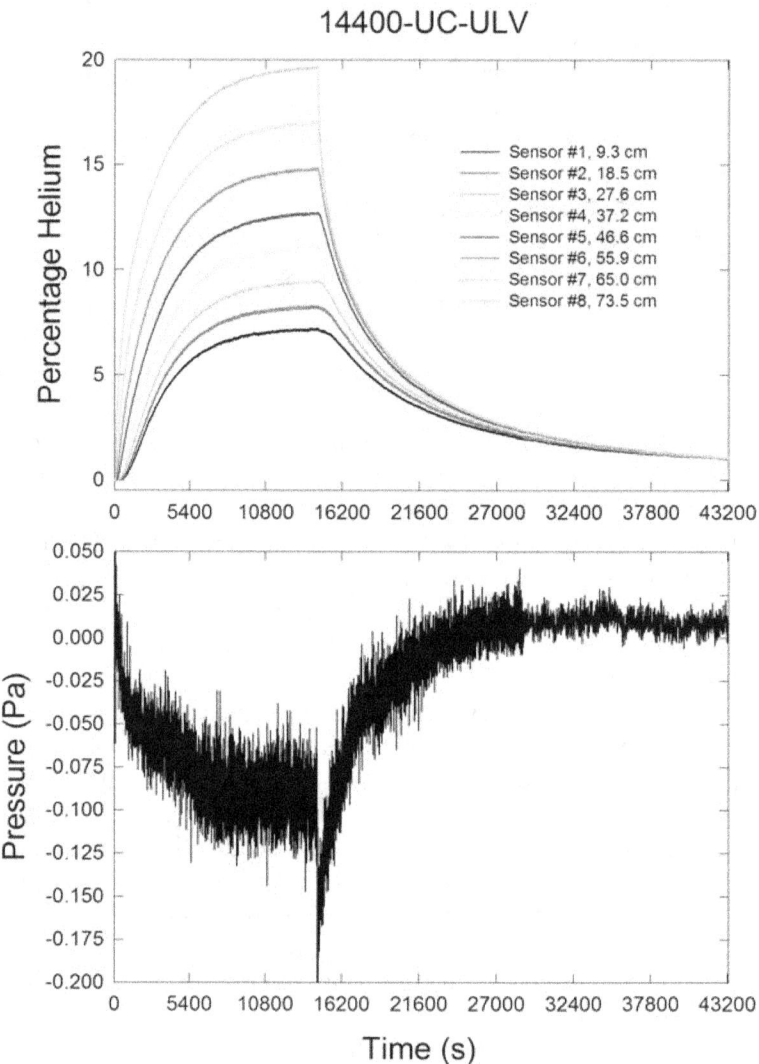

Figure 45. The upper plot shows helium volume percent as a function of time at eight locations for a four hour release of helium from the upper center position into the ¼-scale garage with 2.15 cm × 2.15 cm upper and lower openings in the front face. The lower plot shows the differential pressure across the face.

As observed for other release periods and locations, the times for initial detection of helium following the start of a release were very similar when the two-vent and single large vent walls were used. The periods for the single small vent were considerably longer at lower positions. These results provide additional confirmation that the initial filling behavior is dependent on the total vent area and not the vent locations.

At the end of the helium release period there were substantial vertical concentration gradients when the two-vent wall was used. The values of $t_{V\%max}$ in Table 3 show that the helium concentrations continued to rise for short periods at each sensor location, but after this the helium volume percents began to fall, and the vertical concentration gradient began to decrease. Within an hour the helium concentration in the garage had decreased substantially more than observed for the single vents. The largest relative change was at the upper sensor locations. The helium volume fractions observed for this experiment were the lowest observed during the test series. This likely reflects the lower concentrations present at the end of the release period as compared to those for releases nearer the floor. Comparison of the $V\% Slope_{t=18000}$ values

62

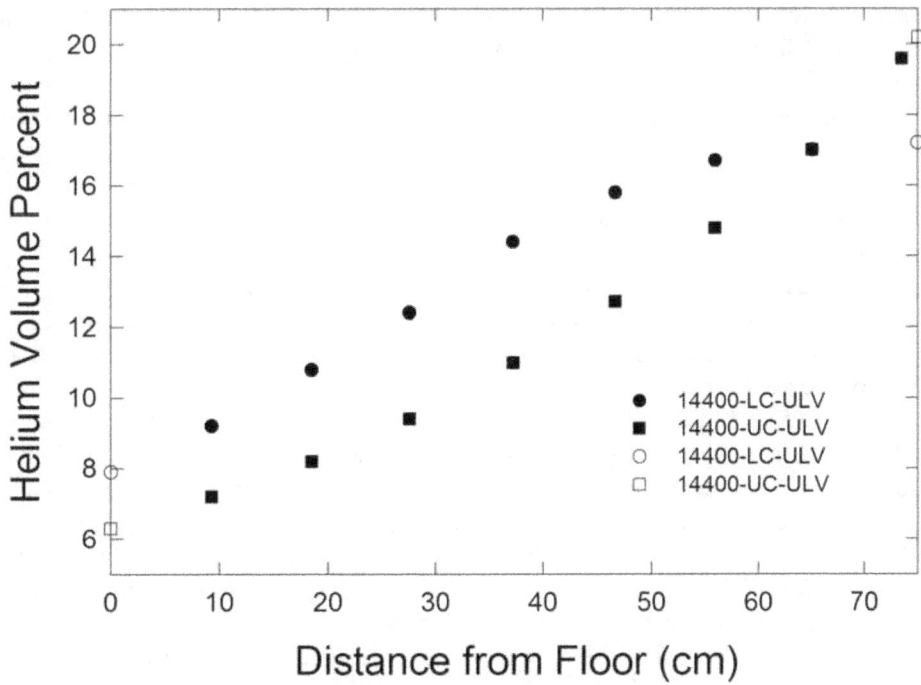

Figure 46. Values of experimental helium volume percents (solid symbols) are plotted as a function of distance above the floor for 14400-LC-ULV (circle) and 14400-UC-ULV (square). Estimated values at the floor and ceiling are indicated by open symbols.

for 14400-LC-ULV and 14400-UC-ULV reveals that the helium loss rates increased with height, and that the rates were substantially higher for the release near the floor. The latter observation reflects the strong dependence of the helium loss rate from the garage on the average concentration within the garage.

The differential pressure data shown in Figure 45 shows that the pressure continuously fell and became negative following an initial pressure rise of $\Delta P = 0.01$ Pa (see Table 4) when the helium flow was initiated. By the end of the four hour release the differential pressure had dropped to -0.09 Pa. When the helium flow was stopped, there was an additional drop of $\Delta P = -0.09$ Pa. The negative pressure difference began to dissipate as helium was lost from the garage during the post-release period and approached zero as the air concentration within the garage neared 100 %. The value of ΔP recorded when the air flow used to sweep helium from the enclosure was shut off was in good agreement with the other experiments in which the two-vent wall was used (see Table 4).

4. Discussion

An extensive series of measurements of concentration vertical profiles and differential pressure measurements have been provided that yield insights into the mixing behavior when buoyant plumes are released into partially enclosed spaces, such as garages, having idealized openings with total areas typical of those found in the real world. Parameters varied include vent size, number, and location, helium release rate, and helium release location.

The findings can be generalized as follow:

1. For sensors well removed from the helium plume, initial helium detection occurred at the highest sensor location with subsequent detection taking place sequentially from higher to lower sensors.

63

2. The periods required for initial detection of helium at the various sensors following the start of a release at a given location increased with the area of the vent(s), but was independent of vent number and location.

3. The periods required for initial detection of helium were reduced near the top of the garage and increased nearer the floor when the helium was released near the ceiling as compared to near the floor.

4. Over most of the volume of the garage helium concentration gradients in horizontal planes were minimal and could be neglected.

5. Changing the release location from the lower center to the lower rear position had minimal effects on helium concentration profiles and time behavior.

6. For the one hour releases, the helium concentrations present at the end of a release were independent of the vent area for the two single vents used.

7. For the four hour releases, the helium concentrations at the end of a release were reduced by small fractions when the single large vent was used compared to experiments with the small vent.

8. Higher fractions of released helium were lost from the garage equipped with a given single vent during four hour releases as compared to one hour releases.

9. Moving the release point from near the floor to near the ceiling substantially modified the vertical concentration distributions present at the end of a release, resulting in higher average vertical helium volume fraction gradients.

10. Moving the release point from the near the floor to near the ceiling did not substantially modify the average concentration of helium at the end of a release in the garage for experiments with single centered vents, but led to a substantial reductions when two vertically separated vents were used.

11. The presence of two, vertically well separated vents substantially reduced the amount of helium within the enclosure and increased the vertical concentration gradients present at the end of the release period, as compared to experiments with single vents.

12. Helium exchange rates with the ambient surroundings during the post-release periods were roughly a factor of two higher for experiments with the large single vent as compared to those with the small single vent (area ratio = 1.6).

13. Helium exchange rates with the ambient surroundings during the post-release periods were many times higher for cases with two vertically separated vents as compared to those with a single centered vent having nominally the same total area.

14. Fits to the fan tests failed to predict accurately the small pressure increases due to gas volume flow rate flows that were much smaller than those used in the fan tests.

15. Differential pressures measured between the inside and outside of the garage equipped with upper and lower vents immediately following the end of a helium release were predicted well by a model that accounted for differences in the hydrostatic pressure profiles inside and outside due to different vertical density profiles and balanced the pressure drops at the vents due the resulting induced flows.

16. In the absence of counteracting processes, molecular diffusion smoothed out vertical concentration gradients in periods on the order of one hour.

Even though the behaviors summarized above are fairly complex, they can be qualitatively understood by considering the competitions between various processes taking place within and exterior to the garage during the experiments. These processes include the flow through an opening induced by a pressure difference across the opening, flows generated by buoyancy differences, the entrainment and mixing behavior of flowing fluid, hydrostatic pressure differences, and mixing due to molecular diffusion.

Consider the case of a helium release into the garage of the type studied here. When the helium flow is initiated near the floor, the released fluid immediately begins to accelerate upward due to the force created by the large density difference between helium and air. The rising helium starts mixing with surrounding fluid, which is pure air when the flow is first started. The mixing is due to several mechanisms. At the molecular level, mixing results from molecular diffusion. The flowing helium generates a low pressure region that pulls air toward and into the upward flowing plume and increases the concentration gradients, resulting in increased molecular mixing. The flow also becomes unstable and develops large-scale vortical motions that can engulf additional ambient fluid and substantially increase the entrainment and mixing rates. Measurements discussed earlier showed that when the initial flow reached a height of 65.0 cm, a flow distance 44.3 cm, the helium concentration had already dropped by a factor of 12. (e.g., see Figure 25).

When the buoyancy-driven flow reaches the ceiling its momentum causes the flow to change direction and begin to spread out over the ceiling in the form of a ceiling jet. A similar flow turning takes place when the ceiling jet reaches the walls. This turning results in a downward flow of the fluid, but in this case the flow is moving into a region of higher density and begins to decelerate and eventually stops. The results of these flow behaviors is that a stable layer of light gas forms near the ceiling and begins to move downward as more and more fluid enters from below. The experimental findings show that over most of the enclosure volume the upper-layer mixture is laterally uniform. The growth of the ceiling layer is therefore similar to the filling of a container with water.

If the flows discussed above were the only processes affecting the upper layer it might be expected that the area would be a well mixed region having a sharp interface with the fluid below. However, there are two mixing processes that tend to mix the upper and lower layers. The first is molecular diffusion, which smears out the sharp interface and, transports fluid from the lower density upper layer into the higher density lower layer. The second mixing process is due to a flow that can develop as a result of entrainment by the buoyant plume. The flow that develops due to the lower density of the released helium lowers the local pressure of the gas within the plume. As a result, a low-speed flow develops that entrains surrounding gas into the buoyant plume. For a buoyancy-induced flow in the open the entrained flow tends to be a horizontal flow into the plume. However, for a partially closed environment with small openings, such as a garage, the entrainment creates pressure forces that lead to the development of a low speed recirculation throughout large volumes of the space that can transport fluid from the upper layer to the lower layer. In a completely closed system the recirculation flow must develop, while in a partially closed system it is possible that the required entrainment flow can be developed by partially or entirely balancing flows into and out of the space at different locations.

The introduction of a gas flow into the partially enclosed space results in a pressure increase within the space. Since the pressure outside of the enclosure is assumed to be unchanged, this requires that pressure differences develop across any openings connecting the inside and outside of the enclosure and, as a result, flows develop through the vents that counteract the pressure buildup due the volume of the introduced gas. Since the flow rates through an opening increase with differential pressure, a balance is eventually reached in which the volume of gas flowing out of the space is matched by a pressure difference between the interior and exterior that equals the pressure drops induced by any flows through the openings. For the cases studied here, the resulting differential pressures were small, and the balancing of pressure and flow was rapid.

The role of hydrostatic pressure differences in generating flows between the interior and exterior of the garage has already been discussed. The differential pressure differences can be substantial when the relevant vertical distances and the density differences between interior and exterior are large. For the wall with two vents studied here, the pressure differences between the interior and exterior due to hydrostatic pressure variations are sufficient to create flows that significantly accelerate the removal of helium from the garage as compared to cases with a single opening having the same nominal open area.

The observation that the average helium concentrations within the garage at the end of a release were unchanged when the single small vent was used for one- and four hour releases indicates that there was no outside air entering the enclosure during the release and that gas exchange with the exterior was unidirectional. The same seems to have been true for the one hour releases with a single large vent. Interestingly, when the release period was increased to four hour for the cases with a single large vent, some air apparently entered the garage during the release period since the average helium concentrations in the garage at the conclusion of the releases were reduced compared to comparable experiments with the single small vent. This suggests that the vent flow induced by the in-flow of helium for the four hour release did not have sufficient velocity to prevent some backflow of air through the large vent. The backflow could have been due to molecular diffusion or the small hydrostatic pressure difference between the interior and exterior expected to develop over the height of the opening.

It is evident that when vents were located near the top and bottom of the front wall that hydrostatic pressure differences developed that were sufficient to overcome the positive pressure associated with the helium flow into the garage, and significant amounts of air flowed into the garage through the bottom vent. The differential pressure increased with time as the helium concentration built up. The near leveling off of the helium concentrations with time for the four hour releases with the two-vent front wall suggests that the volume of the in-flow of air through the lower vent was nearly balanced by the volume flow of the helium/air mixture introduced into the garage.

Very minor differences were observed when the release point near the floor was moved from the center of the garage to a position near the rear wall. This indicates that the entrainment behavior of the buoyancy-induced plume and the development of the stable upper layer were not sensitive to lateral release locations.

Significant changes in the helium vertical concentration distributions were observed when the release point was moved from near the floor to near the ceiling. One reason for this was that the gas flowing into the upper layer had higher helium concentrations since the distance that the buoyancy-induced plume traveled before striking the ceiling, and thus the amounts of air and lower-helium concentration helium/air mixture entrained, was significantly reduced. Since the buoyancy-induced flow was not passing through the lower part of the enclosure, it was no longer necessary for a large-scale recirculation pattern transporting low density fluid from near the ceiling into the lower part of the enclosure to develop. Note that these mechanisms also explain why the periods required for first helium detection were shortened for sensors near the ceiling and increased for sensors near the floor.

Average concentrations in the garage were unchanged with the helium release height for experiments with single vents even though the vertical concentration gradients were very different. This observation suggests that the total amount of helium that flowed from the enclosure during a release due to the induced pressure associated with the helium flow did not change. In turn, this suggests that the average concentration of helium flowing through the exit vent was the same.

In contrast, the average helium concentrations within the garage at the end of a release were significantly reduced when the upper-release location with a two-vent front wall was used as compared to releases near the floor. The likely reason for this is that the helium volume fraction in the layer near the ceiling, and thus in the gas flowing through the upper vent, was higher. As a result more helium flowed from the enclosure, while the air that entered near the floor at the same time did not move upward as quickly.

During the post-release period the pressure differences due to the helium in-flow were no longer present and the buoyancy-driven flow depositing high helium concentrations near the ceiling rapidly died away. As a result, it was to be expected that molecular diffusion would have begun to dissipate the vertical concentration gradients. On the other hand, any air that entered the enclosure sank towards the floor while mixing with the surrounding gas. This process would have tended to lower the helium concentration in the lower part of the enclosure and could have counteracted the effects of molecular diffusion. It was experimentally observed that when single vents were used the concentration gradients present at the end of helium release tended to dissipate fairly rapidly, but that small residual concentration gradients were present for much longer periods. These observations indicate that the rates of air inflow through either of the two sizes of single vent were relatively slow as compared to the time required for molecular diffusion to smooth out the vertical concentration gradients.

The gas exchange rates between the interior and exterior during the post-release period were observed to be roughly two times higher for the large vent as compared to the small vent. If molecular diffusion alone was responsible, it would have been expected that the exchange rates would have depended only on the area ratio of 1.6. The higher ratio is an indication that the higher differential pressure difference associated with the increased height of the larger vent augmented the flow of air into the garage.

When cases with two-vents were compared with those for single vents, gas exchange rates between the interior and exterior were many times higher. Furthermore, the vertical concentration gradients for a given helium concentration level were considerably higher than when single vents were used. These observations demonstrate that the exchanges were dominated by the flows through two vents that developed due to hydrostatic pressure differences between the interior and exterior. Since primarily air entered the vent near the floor and high-concentration helium gas exited near the ceiling, the vertical concentration gradients were maintained.

When the release location was moved from near the floor to near the ceiling, the effects of the hydrostatically-induced pressure differences were enhanced since the helium concentration gradients were larger.

The conditions studied here were extremes. Clearly in the real world, a continuum of conditions is possible. Which of the various physical processes dominate a particular release at a particular time will clearly vary depending on such properties as release rate, release location, total vent area, and vent distribution.

Thus far it has been tacitly assumed that the environmental conditions outside of the garage were constant. Over short periods this was a good assumption. However, some of the experiments lasted several days. Over such long time periods the effects of changes in the environment could be observed. For instance, the exterior pressure changed relatively slowly due to atmospheric pressure fluctuations. These changes induced flows into or from the enclosure (depending on whether the atmospheric pressure was increasing or decreasing). The effects of such changes became evident at long times for experiments (particularly those with a small single vent) where exchange rates between the interior and exterior were otherwise slow. Other small exterior changes in pressure were likely associated with ventilation flows into the laboratory and minor changes in temperature. These properties were not monitored during the experiments, but it is considered likely that they had minor effects on the observed behaviors.

It is important to keep in mind that these experiments were conducted under laboratory conditions where the temperature was well regulated and flow velocities in the ambient were on the order of a few tens of cm/s. As a result, measured differential pressures between the garage interior and the exterior were well below the nominal value of 4 Pa often assumed to be representative of the pressure difference between a building interior and the surrounding ambient. As a result, the observed values of ACHs are expected to be much lower than the corresponding $(ACH)_{4Pa}$s included in Table 2. Actual ACHs with only air inside the enclosure were not determined, but the loss rate of helium from the enclosure with the various types of vents provides a means for estimating the effective ACH for a given condition. As an example, consider the case of a one hour release near the floor with a single small vent in the center of the front wall. The measurements for sensor #4 shown in Figure 19 indicate that the helium volume fraction dropped from 0.391 to 0.369 over the one hour period extending from one to two hours after the end of the release. The concentration fall off should approximately obey the relation

$$\ln \frac{c}{c_o} = -\frac{Q_{enc}}{V_{enc}}(t - t_o) \qquad (10)$$

where c and c_o are the helium volume fractions at time t and t_o, respectively. Substituting the helium volume fractions above yields an ACH value of 0.06 which is substantially less than the value of $(ACH)_{4Pa}$ = 1.98 included in Table 2. This is the case even though it would be expected that helium would be lost from the enclosure faster than if the interior were simply air.

A similar estimate for the one hour release at the center of enclosure near the floor for the enclosure with vents near the floor and ceiling yielded an ACH value of 0.6 for $c = 0.64$ and $c_o = 0.114$. While ten times larger than the value estimated for the single small vent, it remains substantially smaller than the corresponding $(ACH)_{4Pa}$ of 3.42 listed in Table 2.

Another important point to note is that the helium volume flow rates adopted here, which were scaled assuming constant leak rates sufficient to completely empty a tank containing 5 kg of hydrogen in one or four hours, are also highly idealized and may not represent possible scenarios for actual leaks from hydrogen-powered automobiles. The tanks for such vehicles are highly pressurized, and any leak rates comparable to those assumed here would likely change with time as the pressure in the tank decreased. The leak periods assumed are arbitrary, and no particular release scenario has been assumed. Proposed specifications for the hydrogen fuel tanks specify maximum leak rates that are more than 500 times smaller than the rates assumed here. Current tanks in use contain pressure release devices designed to release all of the hydrogen in case the tank becomes overpressurized. Many of these tanks are designed to release 5 kg in approximately one minute, e.g., see [71]. The average volume flow rates for such a release would be 60 to 240 times faster than assumed in the current study, with much higher flow rates at the start of the release.

Interestingly, several experimental studies in the literature designed to characterize hydrogen releases in garages have employed similar volume flow rates to those of 997 L/min and 249 L/min assumed here for a full-scale garage. Gupta et al. considered cases of 5 kg releases of hydrogen lasting from 1.5 hours to 55 hours, corresponding to volume flow rates of 668 L/min and 18 L/min. [14] Lacome et al. studied hydrogen releases having volume flow rates ranging from 690 L/min to 138 L/min. [15] Barley and Gawlik considered leak rates over a range from 2.4 L/min to 50 L/min. [22] Cariteau et al. used volume flow rates of 190 L/min and 569 L/min in their garage investigation. [29]

The general observations concerning the concentration distributions in the garage during releases using single vents are consistent with previous discussions concerning the release of buoyancy-dominated flows inside nearly sealed enclosures. Relevant papers include those of Baines and Turner [72], Germeles [73], Worster and Huppert [74], Cleaver et al. [75], and Kaye and Hunt [76]. These papers generally consider buoyancy-dominated plumes similar to those considered here. Interestingly, they do not typically consider molecular diffusion as a mixing process. The results of the current work suggest that molecular diffusion has an important influence on the observed helium concentration profiles.

In a study concerned with the release of methane within an enclosure Cleaver et al. noted that the methane concentration was uniform across any horizontal section for locations removed from the plume. [75] This is consistent with the conclusion of the current study. Cariteau et al. reported that horizontal concentration gradients existed in the region of a buoyant helium plume within an enclosure, but that the gradients quickly dissipated when the helium flow was halted. [31] Similar results were also observed here. Cariteau et al. attribute the rapid decay to hydrostatic pressure forces in the horizontal direction that are present when there are lateral variations in density. This explanation is likely the reason for the apparent horizontal uniformity of buoyant gas concentrations within enclosures. Whenever a horizontal concentration gradient develops, it creates a hydrostatic pressure variation that induces a horizontal flow that tends to smooth out the variation. Of course, there will be some localized regions where changes in concentration are sufficiently rapid to overcome this tendency to develop uniform horizontal concentration profiles. The buoyant plume has already been discussed. Other such areas are to be expected in the immediate vicinity of vents where air enters the enclosure.

As discussed in Section 1.2, numerous studies have recently appeared which consider the release of hydrogen and/or helium within enclosures. Most of these studies have been motivated by a desire to improve safety as the hydrogen economy develops. A number of these investigations have considered the scenario of a hydrogen-fueled automobile parked within a personal garage. Many of these studies were primarily designed to investigate hydrogen concentration distributions within enclosures in which the vents were small and were simply intended to minimize pressure build up due to the release of gas into the confined space. Some of the studies were designed to investigate the use of relatively large vents to limit the hydrogen build up within the space. Only a limited number of studies have attempted to characterize hydrogen (helium) distribution in model garages with leaks scaled to represent those in real-world garages. Notable among the latter studies are the recent studies designed to characterize a maximum acceptable permeation rate for hydrogen from compressed-gas fuel tanks. [24-30] As discussed in Section 1.3, the minimum assumed *ACH* assumed for these studies was roughly a hundred times smaller than the value adopted for the current study as being representative of garages in the United States.

The data provided in the current investigation is not only unique because leakage areas were scaled to be characteristic of those in actual garages, but also because vent sizes and locations and gas release locations and durations were systematically varied. Another major difference from most earlier investigations is that not only was the development of concentration distributions of released gas during the release period emphasized, but concentration distributions and the loss of the gas from the garage during the post-release period were also characterized. To our knowledge, the current investigation is the first of this type of experiment to report differential pressure measurements and to include "fan test" characterization of the test enclosure. A number of quantitative measures derived from the experimental results are tabulated in Table 3. As a result of the range of variables considered, the well characterized uncertainties, and the tabulation of quantitative measures, the results reported here provide stringent challenges for developing and validating CFD models for predicting the release of buoyant gases inside partially enclosed spaces.

Several groups have reported the use of simplified models for predicting concentration distributions for this type of flow configuration. [19-22,61] Each of these models uses assumptions concerning the vertical concentration distribution in order to calculate the differential pressure resulting from hydrostatic pressure variations between the interior and surroundings of an enclosed space. The differential pressure is then used to predict the induced flow through a vent(s). Since it has been shown here that the measured differential pressure can be accurately predicted based on the observed vertical concentration profiles within the enclosure, the current results should prove particularly important for testing and validating this type of model.

5.   Final Remarks

Detailed results for the sixteen experiments discussed in this report have been posted on the internet in the form of Excel spread sheets. The web addresses for the files and some details concerning individual experiments are included in Appendix A.

These results provide stringent tests for developing and testing CFD codes designed to predict concentration distributions in partially enclosed spaces during and following releases of a buoyant gas for the relatively slow release rates corresponding to scaled one- and four hour releases of 5 kg of hydrogen inside a two-car garage. The data also provide insights into the effects of release location and vent size, number, and location on the concentration distribution that develop and decay during such releases.

While designed to improve the understanding and modeling capability of releases of hydrogen in residential garages, several limitations of the study should be kept in mind. It has been demonstrated that molecular diffusion has a measureable effect on the observed concentration distributions. As a result, the experiments are not similar in the sense that the results for the reduced-scale garage can not be simply scaled to predict the results in a corresponding real-scale garage. The vents studied here are highly idealized. In real garages leaks are expected to occur at multiple, difficult to identify, locations over the entire garage envelop. Gas exchange through the single and double vents considered here may take place in a significantly different manner than when exchange takes place through the narrow leaks present in real garages. Real garages are subject to differential pressure differences associated with impinging wind and temperature differences between the interior and surroundings. Unlike for the current experiments where the laboratory minimizes such effects, the resulting differential pressure differences are expected to either be important or to completely dominate gas exchange between an actual garage and its surroundings, depending on the leak distribution. As a result, in most cases the observed loss rates from the reduced-scale garage are lower limits.

6.   References

1. M. D. Koontz, R. Burruss, D. R. Cade, and H. E. Rector, Hydrogen Emissions from Electric Vehicle Batteries Undergoing Charges in Residential Garages, Electric Power Research Institute Technical Report 103421, Palo Alto, CA, December 1993, 115 pp.

2. M. R. Swain, J. Shriber, and M. N. Swain, Comparisons of hydrogen, natural gas, liquidfied petroleum gas, and gasoline leakage in a residential garage, Energy Fuels **12** (1), 83–89 (1998).

3. M. R. Swain and M. N. Swain, Passive ventilation systems for the safe use of hydrogen, Intl. J. Hydrogen Energy **21** (10), 823–825 (1996).

4. M. R. Swain, E. S. Grilliot, and M. N. Swain, The application of a hydrogen risk assessment method to vented spaces, in Advances in Hydrogen Energy, C. E. Gregoire Padro and F. Lau, eds., Kluwer Academic Publishers, Hingham, MA (2000) 163–173.

5. M. R. Swain, E. S. Grilliot, and M. N. Swain, Experimental verification of a hydrogen risk assessment method, Chemical Health & Safety **6** (3), 28–32 (1999).

6. V. Agranat, Z. Cheng, and A. Tchouvelev, CFD modeling of hydrogen releases and dispersion in hydrogen energy station, in Proceedings of the 15th World Hydrogen Energy Conference, Yokohama, Japan (June 27 – July 2, 2004).

7. M. R. Swain, P. Filoso, E. S. Grilliot, and M. N. Swain, Hydrogen leakage into simple geometric enclosures, Intl. J. Hydrogen Energy **28** (2), 229–248 (2003).

8. M. R. Swain, Addendum to Hydrogen Vehicle Safety Report: Residential Garage Safety Assessment, Directed Technologies, Inc., Arlington, VA, August 1998, 13 pp.

9. E. A. Papanikolaou and A. G. Venetsanos, CFD modeling for helium releases in a private garage without forced ventilation, First International Conference on Hydrogen Safety, Pisa, Italy (September 8-10, 2005).

10. W. Breitung, G. Necker, B. Kaup, and A. Veser, Numerical simulation of hydrogen release in a private garage, in Proceedings of the 4th International Symposium on Hydrogen Power - Theoretical and Engineering Solutions, Strahlsund, Germany (September 9-14, 2001) 368–377.

11. E. Gallego, E. Migoya, J. M. Martin-Valdepenas, A. Crespo, J. Garcia, A. Venetsanos, E. Papanikolaou, S. Kumar, E. Studer, Y. Dagba, T. Jordan, W. Jahn, S. Hoiset, D. Makarov, and J. Piechna, An intercomparison exercise on the capabilities of CFD models to predict distribution and mixing of H_2 in a closed vessel, Intl. J. Hydrogen Energy **32** (13), 2235–2245 (2007).

12. Y. N. Shebeko, V. D. Keller, O. Y. Yeremenko, I. M. Smolin, M. A. Serkin, and A. Y. Korolchenko, Regularities of formation and combustion of local hydrogen-air mixtures in a large volume, Chem. Ind. **24**, 21–47 (1988). (in Russian)

13. S. V. Puzach, Some features of formation of local combustible hydrogen–air mixtures during continuous release of hydrogen in a room, Intl. J. Hydrogen Energy **28** (9), 1019–1026 (2003).

14. S. Gupta, J. Brinster, E. Studer, and I. Tkatschenko, Hydrogen related risks within a private garage: concentration measurements in a realistic full scale experimental facility, Intl. J. Hydrogen Energy **34** (14), 5902–5911 (2009).

15. J. M. Lacome, Y. Dagba, D. Jamois, L. Perrette, and Ch. Proust, Large-scale hydrogen release in an isothermal confined area, Second International Conference on Hydrogen Safety, San Sebastian, Spain (September 11-13, 2007).

16. A. V. Tchouvelev, J. DeVaal, Z. Cheng, R. Corfu, R. Rozek, and C. Lee, CFD modeling of hydrogen dispersion experiments for SAE J2578 test method development, Second International Conference on Hydrogen Safety, San Sebastian, Spain (September 11-13, 2007).

17. Y. Ishimoto, E. Merilo, M. Groethe, S. Chiba, H. Iwabuchi, and K. Sakata, Study of hydrogen diffusion and deflagration in a closed system, Second International Conference on Hydrogen Safety, San Sebastian, Spain (September 11-13, 2007).

18. A. G. Venetsanos, E. Papanikolaou, M. Delichatsios, J. Garcia, O. R. Hansen, M. Heitsch, A. Huser, W. Jahn, T. Jordan, J-M. Lacome, H, S. Ledin, D. Makarov, P. Middha, E. Studer, A. V Tchouvelev, A. Teodorczyk, F. Verbecke, and M. M. van der Voort, An inter-comparison exercise on the

capabilities of CFD models to predict the short and long term distribution and mixing of hydrogen in a garage, Intl. J. Hydrogen Energy **34** (14), 5912–5923 (2009).

19. J. Zhang, J. Hereid, M. Hagen, D. Bakirtzis, M. A. Delichatsios, and A. G. Venetsanos, Numerical studies of dispersion and flammable volume of hydrogen in enclosures, Second International Conference on Hydrogen Safety, San Sebastian, Spain (September 11-13, 2007).

20. J. Zhang, M. A. Delichatsios, and A. G. Venetsanos, Numerical studies of dispersion and flammable volume of hydrogen in enclosures, Intl. J. Hydrogen Energy **35** (12), 6431–6437 (2010).

21. C. D. Barley, K. Gawlik, J. Ohi, and R. Hewett, Analysis of buoyancy-driven ventilation of hydrogen from buildings, Second International Conference on Hydrogen Safety, San Sebastian, Spain (September 11-13, 2007).

22. C. D. Barley and K. Gawlik, Buoyancy-driven ventilation of hydrogen from buildings: laboratory test and model validation, Intl. J. Hydrogen Energy **34** (13), 5592–5603 (2009).

23. B. J. Lowesmith, G. Hankinson, C. Spataru, and M. Stobbart, Gas build-up in a domestic property following releases of methane/hydrogen mixtures, Intl. J. Hydrogen Energy **34** (14), 5932–5939 (2009).

24. P. Adams, A Bengaquer, B. Cariteau, V. Mokkov, and A. G. Venetsanos, Allowable hydrogen permeation rate from road vehicle compressed gaseous storage systems in garages; part 1—introduction, scenarios, and estimation of an allowable permeation rate, Third International Conference on Hydrogen Safety, Ajaccio, France (September 16-18, 2009).

25. A. G. Venetsanos, E. Papankolaou, B. Cariteau, P. Adams, and A. Bengaouer, Estimation of allowable hydrogen permeation rate from road vehicle compressed gaseous storage systems in garages; part 2: CFD dispersion calculations using the ADREA-HF code and experimental validation using helium tests at the garage facility, Third International Conference on Hydrogen Safety, Ajaccio, France (September 16-18, 2009).

26. J.-B Saffers, D. Makarov, and A. V. Molkov, Estimation of allowable hydrogen permeation rate from road vehicle compressed gaseous storage systems in garages; part 3: modelling and numerical simulation of hydrogen-permeation in a garage with adiabatic walls and still air, Third International Conference on Hydrogen Safety, Ajaccio, France (September 16-18, 2009).

27. P. Adams, A. Bengaouer B. Cariteau V. Molkov and A. G. Venetsanos, Allowable hydrogen permeation rate from road vehicles, Intl. J. Hydrogen Energy, 36 (3), 2742–2749 (2011).

28. A. G. Venetsanos, E. Papanikolaou, B. Cariteau P. Adams, and A. Bengaouer, Hydrogen permeation from CGH2 vehicles in garages: CFD dispersion calculations and experimental validation, Intl. J. Hydrogen Energy **35** (8), 3848–3856 (2010).

29. B. Cariteau, J. Brinster, and I. Tkatschenko, Experiments on the distribution of concentration due to buoyant gas low flow rate release in an enclosure, Intl. J. Hydrogen Energy, **36** (3), 2505–2512 (2011).

30. J-B. Saffers, D. Makarov, and V. V. Molkov, Modelling and numerical simulation of permeated hydrogen dispersion in a garage with adiabatic walls and still air, Intl. J. Hydrogen Energy, **36** (3), 2582–2583 (2011).

31. B. Cariteau, J. Brinster, E. Studer, I. Tkatschenko and G. Joncquet, Experimental results on the dispersion of buoyant gas in a full scale garage from a complex source, Intl. J. Hydrogen Energy, **36** (3), 2489–2496 (2011).

32. E. G. Merilo, M. A. Groethe, J. D. Colton, and S. Chiba, Experimental study of hydrogen release accidents in a vehicle garage, in Third International Conference on Hydrogen Safety, Ajaccio, France (September 16–18, 2009).

33. E. G. Merilo, M. A. Groethe, J. D. Colton, and S. Chiba, Experimental study of hydrogen release accidents in a vehicle garage, Intl. J. Hydrogen Energy, **36** (3), 2436–2444 (2011).

34. S. Benteboula, A. Bengaouer, and B. Cariteau, Comparison of two simplified models predictions with experimental measurements for gas release within an enclosure, in Third International Conference on Hydrogen Safety, Ajaccio, France (September 16-18, 2009).

35. V. P. Denisenko, I. A. Kirillov, S. V. Korobtsev, and I. I. Nikolaev, Hydrogen-air explosive envelope behaviour in confined space at different leak velocities, in Third International Conference on Hydrogen Safety, Ajaccio, France (September 16-18, 2009).

36. Y. A. Skob, M. L. Ugryumov, K. P. Korobchynkiy, V. V. Shentsov, E. A. Granovskiy, and V. A. Lyfar, Numerical modeling of hydrogen deflagration dynamics in enclosed space, in Third International Conference on Hydrogen Safety, Ajaccio, France (September 16-18, 2009).

37. E. A. Papanikolaou, A. G. Venetsanos, M. Heitsch, D. Baraldi, A. Huser, J. Pujol, D. Makorov, V. Molkov, J. Garcia, and N. Markatos, HySafe SBEP-V20: numerical studies of release experiments inside a residential garage with passive ventilation, in Third International Conference on Hydrogen Safety, Ajaccio, France (September 16-18, 2009).

38. E. A. Papanikolaou, A. G. Venetsanos, M. Heitsch, D. Baraldi, A. Huser, J. Pujol, J. Garcia, and N. Markatos, HySafe SBEP-V20: numerical studies of release experiments inside a naturally ventilated residential garage, Intl. J. Hydrogen Energy 35 (10), 4747–4757 (2011).

39. A. Friedrich, A. Vesera, G. Sterna, and N. Kotchourkoa, Hyper experiments on catastrophic hydrogen releases inside a fuel cell enclosure, Intl. J. Hydrogen Energy, 36 (3), 2678–2687 (2011).

40. G. M. Cerchiara, N. Mattei, M. Schiavetti, and M. N. Carcassi, Natural and forced ventilation study in an enclosure hosting a fuel cell, Intl. J. Hydrogen Energy, 36 (3), 2478–2488 (2011).

41. E. Papanikolaou, A. G. Venetsanos, G. M. Cerchiara, M. Carcassi, and N. Markatos, CFD simulations on small hydrogen releases inside a ventilated facility and assessment of ventilation efficiency, Intl. J. Hydrogen Energy, 36 (3), 2597–2605 (2011).

42. A. G. Venetsanos, P. Adams, I. Azkarate, A. Bengaouer, L. Brett, M. N. Carcassi, A. Engebø, E. Gallego, A. I. Gavrikov, O. R. Hansen, S. Hawksworth, T. Jordan, A. Kessler, S. Kumar, V. Molkov, S. Nilsen, E. Reinecke, M. Stöcklin, U. Schmidtchen, A. Teodorczyk, D. Tigreat, and N. H. A. Versloot, On the use of hydrogen in confined spaces: results from the internal project InsHyde, Intl. J. Hydrogen Energy, 36 (3), 2693–2699 (2011).

43. W. M. Pitts, K. Prasad, J. C. Yang, and M. G. Fernandez, Experimental characterization and modeling of helium dispersion in a ¼-scale two-car residential garage, in Third International Conference on Hydrogen Safety, Ajaccio, France (September 16-18, 2009).

44. P. Middha, O. Hansen, and I. E. Storvik, Validation of CFD-model for hydrogen dispersion, J. Loss Prev. Process Industries 22 (6), 1034–1038 (2009).

45. K. Matsuura, H. Kanayama, H. Tsukikawa, and M. Inoue, Numerical simulation of leaking hydrogen dispersion behavior in a partially open space, Intl. J. Hydrogen Energy 33 (1), 240–247 (2008).

46. K. Matsuura, M. Nakano, and J. Ishimoto, The sensing-based adaptive risk mitigation of leaking hydrogen in a partially open space, Intl. J. Hydrogen Energy 34 (20), 8770–8782 (2009).

47. K. Matsuura, Effects of the geometrical configuration of a ventilation system on leaking hydrogen dispersion and accumulation, Intl. J. Hydrogen Energy 34 (24), 9869–9878 (2009).

48. K. Matsuura, M, Nakano, and J. Ishimoto, Forced ventilation for sensing-based risk mitigation of leaking hydrogen in a partially open space, Intl. J. Hydrogen Energy 35 (10), 4776–4786 (2010).

49. M. R. Swain, E. S. Grilliot, and M. N. Swain, Risks incurred by hydrogen escaping from containers and conduits, Proceedings of the 1998 US DOE Hydrogen Program Review, NREL/CP-570-25315 (1998).

50. K. Matsuura, H. Kanayama, H. Tsukikawa, and M. Inoue, Researches on hydrogen dispersion behavior in a partially open space, J. Hydrogen Energy Syst. Soc Japan 31 (2), 50–57 (2006). [in Japanese].

51. H. Liu and W. Schreiber, The effect of ventilation system design on hydrogen dispersion in a sedan, Intl. J. Hydrogen Energy 33 (19), 5115–5119 (2008).

52. Y. Kim, J. H. Nam, D. Shin, T.-Y. Chung, and Y.-G. Kim, Computational fluid dynamics simulations for hydrogen dispersion and exhaust in residential fuel cell systems, Current Appl. Phys. **10** (Sp. Iss. SI Suppl. 2), S81–S85 (2010).

53. B.-G. Kim, J. S. Kim, and C. Lee, Lagrangian stochastic model for buoyant gas dispersion in a simple geometric chamber, J. Loss Prev. Process Industries **22** (6), 995–1002 (2009).

54. S. K. Vudumu and U. O. Koylu, Detailed simulations of the transient hydrogen mixing, leakage and flammability in air in simple geometries, Intl. J. Hydrogen Energy **34** (6), 2824–2833 (2009).

55. M. F. El-Amin and H. Kanayama, Boundary layer theory approach to the concentration layer adjacent to a ceiling wall at impinging region of a hydrogen leakage, Intl. J. Hydrogen Energy **33** (21), 6393–6400 (2008).

56. M. F. El-Amin, M. Inoue, and H. Kanayama, Boundary layer theory approach to the concentration layer adjacent to the ceiling wall of a hydrogen leakage: axisymmetric impinging and far regions, Intl. J. Hydrogen Energy **33** (24), 7642–7647 (2008).

57. M. F. El-Amin and H. Kanayama, Boundary layer theory approach to the concentration layer adjacent to a ceiling wall at impinging region of a hydrogen leakage, Intl. J. Hydrogen Energy **34** (3), 1620–1626 (2009).

58. K. McGrattan, H. Baum, R. Rehm, W. Mell, R. McDermott, S. Hostikka, and J. Floyd, Fire dynamics simulator (version 5) technical reference guide. NIST Special Publication SP 1018-5, Version 5.5, National Institute of Standards and Technology, Gaithersburg, MD, October, 2010.

59. K. Prasad, N. Bryner, M. Bundy, T. Cleary, A. Hamins, N. Marsh, W. Pitts, and J. Yang, Numerical simulation of hydrogen leakage and mixing in large confined spaces, in 2008 Proceedings of the NHA Annual Hydrogen Conference & Expo, National Hydrogen Association, Sacramento, CA (March 30-April 3, 2008).

60. K. Prasad, W. Pitts, and J. Yang, A numerical study of hydrogen or helium release and mixing in partially confined space, in 2009 Proceedings of the NHA Annual Hydrogen Conference & Expo, National Hydrogen Association, Columbia, SC (March 30-April 3, 2009).

61. K. Prasad, W. Pitts, and J. Yang, Effect of wind and buoyancy on hydrogen release and dispersion in a compartment with vents at multiple levels, Intl. J. Hydrogen Energy **35** (17), 9218–9231 (2010).

62. M. H. Sherman and R. Chan, Building Airtightness: Research and Practice, Report LBNL-53356, Lawrence Berkeley National Laboratory, Livermore, CA, 2003.

63. S. J. Emmerich, J. E. Gorfain, M. Huang, and C. Howard-Reed, Air and Pollutant Transport from Attached Garages to Residential Living Spaces, Internal Report 7072, National Institute of Standards and Technology, Gaithersburg, MD, December 2003.

64. Ventilation for Acceptable Indoor Air Quality, ANSI/ASHRAE Standard 62-1989, American Society of Heating, Refrigerating, and Air Conditioning Engineers, Atlanta, GA, 1990.

65. International Mechanical Code, International Code Council, Inc., Country Club Hills, IL, 2009.

66. G. Mulholland, and M. Fernandez, Report on the Calibration of the Gilibrator-2 Soap Film Flowmeter, Unpublished Report, National Institute of Standards and Technology, February 24, 1997.

67. C. J. Chen and W. Rodi, Vertical Turbulent Buoyant Jets—A Review of Experimental Data, 1980, Pergamon Press, Oxford.

68. W. M. Pitts, G. W. Mulholland, B. D. Breuel, E. L. Johnsson, S. Chung, R., Harris, and D. E. Hess, Real-Time Suppressant Concentration Measurement, in R. G. Gann, ed., Fire Suppression System Performance of Alternative Agents in Aircraft Engine and Dry Bay Laboratory Simulations, Vol. II, Special Publication 890, National Institute of Standards and Technology, pp. 319-585, 1995.

69. V. P. Denisenko, I. A Kirillov, S. V. Korobtsev, I. I. Nikolaev, A. V. Kuznetsov, V. A. Feldstein and V.V. Ustinov, Hydrogen Subsonic Upward Release and Dispersion Experiments in Closed Cylindrical

Vessel, Second International Conference on Hydrogen Safety, San Sebastian, Spain (September 11-13, 2007).

70. Y. S. Touloukian, P. E. Lily, and S. C. Saxena, Thermophysical Properties of Matter, Volume 3, Thermal Conductivity Nonmetallic Liquids and Gases, IFI/Plenum, New York (1970).

71. A. G. Venetsanos, E. Papanikolaou, O. R. Hansen, P. Middha, J. Garcia, M. Heitsch, D. Baraldi, and P. Adams, HySafe standard benchmark Problem SBEP-V11: predictions of hydrogen release and dispersion from a CGH2 bus in an underpass, Intl. J. Hydrogen Energy **35** (8), 3857–3867 (2010).

72. W. D. Baines and J. S. Turner, Turbulent buoyant convection from a source in a confined region, J. Fluid Mech. **37** (1), 51–80 (1969).

73. A. E. Germeles, Forced plumes and mixing of liquids in tanks, J. Fluid Mech. **71** (3), 601–623 (1975).

74. M. G. Worster and H. E. Huppert, Time-dependent density profiles in a filling box, J. Fluid Mech. **132**, 457–466 (1983).

75. R. P. Cleaver, M. R. Marshal, and P. F. Linden, The build-up of concentration within a single enclosed volume following a release of natural gas, J. Hazardous Materials **36** (3), 209–226 (1994).

76. N. B. Kaye and G. R. Hunt, Time-dependent flows in an emptying filling box, J. Fluid Mech. **520**, 135–156 (2004).

Appendix A—Downloadable Data Files of Experimental Results for the Eighteen Tests

The measured helium volume fractions (reported as percentages) at eight locations and differential pressure in Pascals as a function of time in seconds are available on the internet for the eighteen experiments discussed in this report. The results for each experiment are stored as individual Microsoft Excel files that are collected in a single zip file located at
http://www.nist.gov/el/fire_protection/buildings/upload/HeliumDispersionDataSets.zip.

The file names for the experiments use the naming convention described on page 23 of this report. The data are organized as follows:

Column A	Column B	Column C	Column D	Column E	Column F	Column G	Column H	Column I	Column J
Time	Sensor #1	Sensor #2	Sensor #3	Sensor #4	Sensor #5	Sensor #6	Sensor #7	Sensor #8	Different. Pressure

In the following sections, the location of sensor #8 within the enclosure is provided along with any special notes for each set of data. The locations for sensors #1 to #7, which did not change, are listed in Table 1 of this report. For convenience, the coordinates of the center of the helium release locations are included below. The sizes of vents in the front face are listed in Table 2. The single vents were centered in the face, while the dual vents were centered horizontally and located 2.54 cm above and below the floor and ceiling, respectively. The nominal periods of the data records are 9 hours following the start of the helium flow, which was initiated 60 s after data collection began.

A1   3600-LC-SSV, One Hour Helium Release near the Floor at the Center of the Garage with a Single 2.40 cm Square Vent

Sensor #8 location: $(x,y,z) = (141$ cm, 134.5 cm, 9.3 cm$)$

Helium release location: $(x,y,z) = (75$ cm, 75, cm, 20.7 cm$)$

A2   3600-LC-SLV, One Hour Helium Release near the Floor at the Center of the Garage with a Single 3.05 cm Square Vent

Sensor #8 location: $(x,y,z) = (141$ cm, 134.5 cm, 9.3 cm$)$

Helium release location: $(x,y,z) = (75$ cm, 75, cm, 20.7 cm$)$

The differential pressure measurement was not recorded during this experiment.

The initial 35 s of the scan for sensor #1 was not recorded.

A3   3600-LC-ULV, One Hour Helium Release near the Floor at the Center of the Garage with Dual 2.15 cm Square Vents

Sensor #8 location: $(x,y,z) = (75.0$ cm, 75.0 cm, 65.0 cm$)$

Helium release location: $(x,y,z) = (75$ cm, 75, cm, 20.7 cm$)$

A4   **3600-LR-SSV, One Hour Helium Release near the Floor at the Rear of the Garage with a Single 2.40 cm Square Vent**

Sensor #8 location: (x,y,z) = (141 cm, 134.5 cm, 9.3 cm)

Helium release location: (x,y,z) = (75 cm, 145, cm, 20.7 cm)

A5   **3600-LR-SLV, One Hour Helium Release near the Floor at the Rear of the Garage with a Single 3.05 cm Square Vent**

Sensor #8 location: (x,y,z) = (141 cm, 134.5 cm, 9.3 cm)

Helium release location: (x,y,z) = (75 cm, 145, cm, 20.7 cm)

Data collection was unexpectedly terminated 2148 s after the helium flow was shut off.

A6   **3600-LR-ULV, One Hour Helium Release near the Floor at the Rear of the Garage with Dual 2.15 cm Square Vents**

Sensor #8 location: (x,y,z) = (75.0 cm, 75.0 cm, 65.0 cm)

Helium release location: (x,y,z) = (75 cm, 145, cm, 20.7 cm)

A7   **3600-UC-SSV, One Hour Helium Release near the Ceiling at the Center of the Garage with a Single 2.40 cm Square Vent**

Sensor #8 location: (x,y,z) = (141 cm, 134.5 cm, 9.3 cm)

Helium release location: (x,y,z) = (75 cm, 75, cm, 72.5 cm)

An air flow into the enclosure was started 22739 s after the helium flow was shut off.

A8   **3600-UC-SLV, One Hour Helium Release near the Ceiling at the Center of the Garage with a Single 3.05 cm Square Vent**

Sensor #8 location: (x,y,z) = (141 cm, 134.5 cm, 9.3 cm)

Helium release location: (x,y,z) = (75 cm, 75, cm, 72.5 cm)

A9   **3600-UC-ULV, One Hour Helium Release near the Ceiling at the Center of the Garage with Dual 2.15 cm Square Vents**

Sensor #8 location: (x,y,z) = (112.5 cm, 37.5 cm, 73.5 cm)

Helium release location: (x,y,z) = (75 cm, 75, cm, 72.5 cm)

A10   14400-LC-SSV, Four Hour Helium Release near the Floor at the Center of the Garage with a Single 2.40 cm Square Vent

Sensor #8 location: (x,y,z) = (141 cm, 134.5 cm, 9.3 cm)

Helium release location: (x,y,z) = (75 cm, 75, cm, 20.7 cm)

A11   14400-LC-SLV, Four Hour Helium Release near the Floor at the Center of the Garage with a Single 3.05 cm Square Vent

Sensor #8 location: (x,y,z) = (141 cm, 134.5 cm, 9.3 cm)

Helium release location: (x,y,z) = (75 cm, 75, cm, 20.7 cm)

Differential pressure measurements for the initial 2462 s are unavailable.

A12   14400-LC-ULV, Four Hour Helium Release near the Floor at the Center of the Garage with Dual 2.15 cm Square Vents

Sensor #8 location: (x,y,z) = (75 cm, 75 cm, 65 cm)

Helium release location: (x,y,z) = (75 cm, 75, cm, 20.7 cm)

A13   14400-LR-SSV, Four Hour Helium Release near the Floor at the Rear of the Garage with a Single 2.40 cm Square Vent

Sensor #8 location: (x,y,x) = (141 cm, 134.5 cm, 9.3 cm)

Helium release location: (x,y,z) = (75 cm, 145, cm, 20.7 cm)

A14   14400-LR-SLV, Four Hour Helium Release near the Floor at the Rear of the Garage with a Single 3.05 cm Square Vent

Sensor #8 location: (x,y,z) = (141 cm, 134.5 cm, 9.3 cm)

Helium release location: (x,y,z) = (75 cm, 145, cm, 20.7 cm)

A15   14400-LR-ULV, Four Hour Helium Release near the Floor at the Rear of the Garage with Dual 2.15 cm Square Vents

Sensor #8 location: (x,y,z) = (75.0 cm, 75.0 cm, 65.0 cm)

Helium release location: (x,y,z) = (75 cm, 145, cm, 20.7 cm)

A16   **14400-UC-SSV, Four Hour Helium Release near the Ceiling at the Center of the Garage with a Single 2.40 cm Square Vent**

Sensor #8 location: (x,y,z) = (141 cm, 134.5 cm, 9.3 cm)

Helium release location: (x,y,z) = (75 cm, 75, cm, 72.5 cm)

A17   **14400-UC-SLV, Four Hour Helium Release near the Ceiling at the Center of the Garage with a Single 3.05 cm Square Vent**

Sensor #8 location: (x,y,z) = (141 cm, 134.5 cm, 9.3 cm)

Helium release location: (x,y,z) = (75 cm, 75, cm, 72.5 cm)

An air flow into the enclosure was started 18660 s after the helium flow was shut off.

A18   **14400-UC-ULV, Four Hour Helium Release near the Ceiling at the Center of the Garage with Dual 2.15 cm Square Vents**

Sensor #8 location: (x,y,z) = (112.5 cm, 37.5 cm, 73.5 cm)

Helium release location: (x,y,z) = (75 cm, 75, cm, 72.5 cm)

www.ingramcontent.com/pod-product-compliance
Lightning Source LLC
Chambersburg PA
CBHW081834170526
45167CB00007B/2803